The Soul Genome

OTHER BOOKS BY PAUL VON WARD

GODS, GENES, AND CONSCIOUSNESS:
Nonhuman intervention in human history

OUR SOLARIAN LEGACY:
Multidimensional humans in a self-learning universe

DISMANTLING THE PYRAMID:
Government by the People

THE SOUL GENOME
Science and Reincarnation

Paul Von Ward

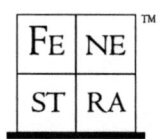

FENESTRA™

The Soul Genome: Science and Reincarnation

Copyright © 2008 Paul Von Ward. All rights reserved. No part of this book may be reproduced or retransmitted in any form or by any means without the written permission of the publisher.

Cover design by Lorin Kee
Cover art licensed by iStock.com
Background photographs on cover provided by NASA

Published by Fenestra Books®
610 East Delano Street, Suite 104
Tucson, Arizona 85705 U.S.A.
www.fenestrabooks.com

Publisher's Cataloging-In-Publication Data
(Prepared by The Donohue Group, Inc.)

Von Ward, Paul, 1939–
 The soul genome : science and reincarnation / Paul Von Ward.

 p. : ill. ; cm.

 Includes bibliographical references and index.
 ISBN: 978-1-58736-995-7

 1. Reincarnation. 2. Soul. 3. Consciousness. I. Title.

BL515 .V659 2008
133.9/01/35   2007940113

If the reincarnation hypothesis withstands the test of 21st-century science, all humans will come to have a different perspective on death. It will be seen as only a stage on a transcendent journey of individual consciousness in its natural process of self-development.

Members of our families who precede us in exiting this life will be honored for their contributions to our own learning and celebrated for leading the way in this journey of conscious evolution. Motivated by this worldview, I wish to celebrate these members of my biological and marriage family-cohort who have already made that transition. This book is dedicated to:

Silo, Buren, Lewis, Nell, Daniel, Gordon, Earl, Dolores, Derrill, and Gloria.

Contents

Preface ix
Introduction 1

Part I
Unsolved Mysteries

1 Prodigies and Precocity 13
2 Anomalous Knowledge 19
3 Dreams and Aspirations 26
4 Past-Life Healing 33
5 Dead Doppelgangers 37
6 Lives That Mimic 42

Part II
Developing Hypothesis

7   A Seminal Experiment 55
8   Self-Learning Universe 62
9   Psychoplasm as Mechanism 70
10 Genotype Entanglements 83
11 Personality Factors 94
12 Sources and Methods 109

Part III
Evaluating Evidence

13 Physical Phenotypes 125
14 Cognitive Cerebrotypes 137
15 Emotional Egotypes 149
16 Social Personatypes 160
17 Creative Performatypes 170
18 Coincidence through Memory 181
19 Cohorts in Consciousness 194

Acknowledgments 207
Appendix 1 209
Appendix 2 210
Appendix 3 211
Appendix 4 212
Appendix 5 213
Endnotes. 215
Bibliography 225
Index. 227
About the Author 233

Preface

Compelling stories of reincarnation can be found in cultural histories around the world. The most widely known currently is that of the Dalai Lama, identified by monks at about age three as the reincarnation of a previous Lama. Given the political importance of the next Dalai Lama, the secular Chinese government preemptively declared it must "approve" his reincarnation.

Are such beliefs in reincarnation caused by a genetic glitch in the human brain? Are they cultural delusions, motivated by a wish for another chance at life? Do they reflect a reasonable interpretation of anecdotes collected over thousands of years? Can those stories and more recently developed evidence lead to a scientific hypothesis that can be validated or proven false?

In my research, I encounter the duelling worldviews that fragment today's society. One view is that because science has no way to prove that an otherworldly process might be false, the possibility that it might be true should not be taken seriously. An opposing view is that it can ultimately be confirmed only by an otherworldly source that a human claims to interpret.

While I recognize its illusive nature, I believe that any subject with as much smoke as reincarnation deserves a scientific search for the fire. Aware of the feelings of believers and skeptics, I attempt to maintain a critical, but open-minded approach to evaluation of a range of evidence. While not dismissing any information that might contain clues to solving the mystery of reincarnation, this project focuses on verifiable information that can be tested by objective means or, at least, evaluations by independent observers.

I suspect that many with science-based views do not want to consider reincarnation because it suggests that a physical, single lifetime may not be all there is to human existence. To consider a one percent probability that consciousness transcends the human brain threatens that model of reality. It also threatens religions based on a single-lifetime view of divine redemption.

People who believe in reincarnation, but as a spiritual function where their god or spirits control the process, are also disturbed. My research poses a challenge to some of their favorite ideas about how reincarnation works. Its requirement to focus on the way nature operates within predictable laws frustrates people who need only metaphysical concepts to buttress their beliefs.

I struggle to find the middle path with a hypothesis where even the use of that term is subject to dispute. The original Greek word *hypotithenai* implied simply placing a supposition or proposition under intellectual scrutiny. To proffer a scientific hypothesis today requires that one define a way in which, at least in principle, it could be proven false. However, by definition, the existence of a "soul" cannot be verified by scientific tools working only within 4-D space-time.

In my effort to take a scientific approach to an "unscientific" concept, I want to make a basic distinction from the start. I accept a colleague's view that the general theory of reincarnation is a meta-theory that cannot, even in principle, be overthrown by the logic of today's science.

I do not expect in this book to prove or disprove a grand reincarnation hypothesis, even in the original Greek sense of the term. We do not yet have the scientific underpinnings necessary to formulate a overall hypothesis of reincarnation. In this respect, reincarnation research is at the stage the black-hole hypothesis was at in physics almost seventy years ago. It took another forty years before the essentials of the black hole theory were in place. And the debate still rages about its falsifiability—whether it can be disproved or not.

My goal is more limited. I attempt to build a framework for collecting and making sense of the smoke—the many types of evidence referred to in the opening paragraph—in order to infer what kind of fire might be responsible. Various mini-hypotheses have been offered to account for the generally accepted evidence. In this book, I offer such a mini-hypothesis—referred to as an Integral Model—as one possible explanation of the tangible or measurable evidence that has given rise to the global concept of reincarnation. I try to ground it in emerging fields of science.

With a few exceptions, books on this subject rely on information from extradimensional sources: psychic readings, intuition, trance-channel messages, spontaneous "recall," or hypnotic regressions. Due to the subjective character of such material, readers cannot verify its veracity. While some of it may be valid, an independent party cannot determine what is and what isn't.

This book is built around the assumption that natural mechanisms must account for the "evidence" that has resulted in a widespread belief in individual, sequential rebirths. I use the term hypothesis to refer to my concept of a mechanism that I believe can account for much of that evidence. The theoretical mechanism I propose can be replicated and evaluated by others. It is based on data that can be understood and tested by independent parties.

Trying to set aside the biases of my own worldview, I started with the raw information that has interested many other researchers. After I examined and categorized various kinds of evidence that cannot be satisfactorily explained by current genetic and personality development theories, I designed and tested a theoretical model that appears to account for the evidence.

Its structure and process may not satisfy some who think it does not meet all conditions of the scientific method. However, for the reasons discussed in the Introduction and Part II, I believe that it is a comprehensive, plausible explanation for all the major areas of evidence.

I incorporated in the model steps to deal with the possibility that much of the evidence could be attributed to chance. I then asked a number of people to join me in testing its reliability and predictability with their own life histories. This book reports on the first phase of that experiment and invites the reader to take part in the continuing search for verifiable evidence that can be used to test my integral reincarnation model against various alternative hypotheses.

You may be surprised to learn that many verified life histories cannot be as logically explained by other theories as they can by a "general reincarnation hypothesis." You may also be amazed to know that if it is real for people like the Dalai Lama, it is equally likely that you and all other humans are reincarnations of people who have lived before. Even giving consideration to the possibility of reincarnation may change the way you think about human behavior.

Just to contemplate that most of what you are today might have come from knowledge and experience gained in many lifetimes may shock you. Consider that whom you marry, or not, what you study in school or college, where you live and work, how you spend your free time, who your friends are, and what you feel about it may reflect the influence of events in centuries past. What difference would it make if you learned that how you interpret global, national, neighborhood, and family affairs may be based on more than what you have learned since birth?

Much thought-provoking evidence suggests that your physical appearance, the way you think, how you react emotionally to life events, the way you interact with other people, and the creative activities and vocations you choose may be predisposed by the experiences of one or more humans who lived in the past. Even if you don't know who they were, you may find what appears to be their "soulprints" in the person you are today and the manner in which you live.

Why do I suggest these radical possibilities? The facts are that credible researchers have thousands of cases where people recall or intuitively act on knowledge and traits that seem to come directly from the private lives of individuals who lived before they were born. When other hypotheses attempt to explain this irrefutable evidence, I believe they cannot measure up to some as-yet-unknown form of linear reincarnation. By "reincarnation" I mean the sequential "carry-forward" of all this inherited data from lifetime to lifetime in some kind of a nonmaterial "it."

The vedic Hindus called it *atman*. The classic Greek and Roman word for it was *psyche*. *Seula* was the Old High German word for it, similar to the Old Norse *sala* and the Gothic *saiwala*. In Old English it was called *sawol*. It may also be seen as analogous to an individual personality or local consciousness.

Regardless of the name one gives the phenomenon, we must deal with the questions of whether and how it fits into the natural universe as we now understand it. For reasons discussed later, I use the Greek term *psychoplasm* in this book. But, for the reader's convenience, I use that term interchangeably with the modern English *soul*. I employ the word soul for its Anglo-Saxon etymology, not because of any supernatural or religious connotation.

The meaning of soul used here is quite different from than generally implied in theology. *In this book, soul or psychoplasm means a genome-like, energetic and information biofield that embodies a single being's knowledge, feelings, and behavior patterns that transcend space-time.* If it can be validated, the psychoplasm as a concept has the potential to reconcile the physical and psychical dimensions just as Einstein's concept of special relativity appeared to reconcile Newton's law of motion with the electromagnetic field.

As Einstein's concept of special relativity is being subjected to revision in the light of new evidence, the psychoplasm hypothesis will need to be revised on the basis of new discoveries. Nevertheless, I hope the tentative findings of this book will stimulate further research on natural and testable explanations for the evidence associated with reincarnation stories. If the existence of a mechanism along the lines of the psychoplasm described here is sustained by more research and experimentation, the study of human consciousness may have fully entered the quantum era.

Introduction

People who want to learn more about reincarnation, perhaps intrigued by a few sensational stories, must contend with as many different theories as there are books. Each has its own set of implicit, usually metaphysical, assumptions. Dramatic claims are made about the why, what, and how of reincarnation with no apparent basis in evidence that readers can assess for themselves.

This state of affairs exists because no consensus on basic terms—much less a theoretical model—has evolved to facilitate public discussion. Aside from psychiatrist Ian Stevenson's legacy (1918–2007) of meticulous documentation of evidence and the ongoing work of his colleagues at the University of Virginia, the subject has received little attention from mainstream scholars and scientists. With mainly metaphysical or anecdotal treatments, most books dealing with the subject generally appeal to audiences who already share the authors' starting premises.

With only a casual interest in the notion of reincarnation and its possible implications, I became acquainted with Stevenson and his work in 1992—we were both speakers at the annual conference of the International Association for New Science in Ft. Collins, Colorado. After further discussion over lunch, I was compelled to read his book book *Twenty Cases Suggestive of Reincarnation*. His lecture and book brought this heretofore—in my own mind—spiritual concept into the world of everyday lives. This perspective resonated with my training in psychology.

Based on decades of research that produced solid, verifiable evidence, Stevenson wrote "... that reincarnation is the best explanation for the stronger cases..."[1] He went on to say that it is also the best explanation for many other cases, but does not claim it is the only explanation. His tradition is being continued and further developed by psychiatrist Jim B. Tucker, M.D.

This book draws from two epistemologies (the scientific approach used by Tucker and his colleagues at the University of Virginia and extradimensional methods of gaining knowledge) without totally ceding the field to either. For that reason it may satisfy neither conventional scientists nor spiritual groups. It uses methods associated with science to temper the imaginative tendencies of mysticism and uses metaphysical concepts to widen the lens of science.

A Natural Science Approach

What I call *natural science* synthesizes Aristotle's method of intellectual inquiry (384–322 BCE) with the approach used by Galileo (1564–1642 CE). The culture known as the dark ages (400–1000 CE) or middle ages (450–1450 CE) in Western Civilization rewarded inquiry that started with *a priori* theological assumptions. However, Aristotle's ideas survived among some Christian, Muslim, and Jewish scholars during the late middle ages. Known as *natural philosophy*, it started with assumptions (axioms) drawn from nature to arrive at laws (theorems).

Galileo and others turned this deductive approach—and theological assumptions—upside down. If one starts with false assumptions, even logical conclusions must be false. Galileo's inductive approach produced three centuries of material progress that overshadowed natural philosophy. Scientists gathered the facts, developed hypotheses, and conducted experiments to test them. With the proofs in hand, technology advanced in leaps and bounds.

However, as physics began to deal with concepts like relativity and quanta in the early twentieth century, science had to make use of both approaches. To make sense of anomalies, deductions based on observations made at a distance were required to account for mounting data about subatomic particles and the behavior of bodies in space. Biology and medicine had to advance deductive hypotheses that could be tested only through statistical probabilities.

This synthesis of natural philosophy and science is natural science. When something cannot be tested, reasonable inferences and assumptions are made on the basis of interpolations and extrapolations from observed data. Until hands-on proofs are available, the assumptions can be formulated as working hypotheses to be tested with calculations of probabilities (logical or mathematical). Scientists then focus their work for a period on the hypothesis which has a higher probability of accounting for the observed evidence. This is the story of the theory of black holes.

In the 1930s, physicists observed happenings in space not predicted by Newton's or Einstein's views of gravity and light. Two scientists introduced the idea that became known as black holes. Of course we had no spacecraft and astronauts to go check them out. That was fortunate because, from what we know now, the humans and their vehicles would have disappeared across a boundary with no escape, unable to communicate what they found.

Scientists were left with observations of activities in space to figure out the cause of the unexpected evidence. They measured the behavior of light, the force of gravity on nearby stars, emissions of X-rays, and the disappearance of objects around the area of a supposed black hole. After forty years of research, a consensus developed that black holes were real even though we could not definitively prove the theory. And, it is still a matter of scientific controversy.

It has been suggested that these "probable" black holes give birth to new universes with the energy they consume from this one. Obviously we cannot prove that hypothesis either. It has been suggested one way to prove it is to find more alleged black holes in our own universe. If we don't find a lot of black holes, according to this view, we will have disproved the theory.

In this context, I believe any hypotheses to explain the presently unexplainable evidence for reincarnation must follow the route of back hole theory. Researchers must prepare themselves to deal only in the realm of probabilities for quite some time to come.

Based on the logic of probability theory, billions of dollars have been invested in the U.S. and Europe on huge nuclear laboratories and particle accelerators in efforts to prove the existence of the illusive Higgs particle and other subatomic phenomena. An Italian physicist, reflecting the view of his colleagues, said, "It is exciting even if you think the chances of it being true are only zero or ten percent."

If less than a ten percent probability exists that reincarnation is a real phenomenon, society should assign it at least a fraction of the billions it allocates to research in physics, psychology, and genetics. Discoveries about reincarnation may be as relevant to our development and species survival as any of those three areas. To begin, we need an initial hypothesis pointing toward a later general reincarnation theory that can be evaluated by the tools of modern science.

I believe the deductive approach used in this book can initially provide that testable hypothesis. If this new hypothesis reliably accounts for the data, it should also predict the presence of similar evidence in proposed new cases. I hope a demonstration of its predictive ability will lead to further research on it and the larger question of reincarnation theory.

Character Vignettes

An abstract discussion of reincarnation cannot hold our attention for very long unless we see issues being played out in lives like our own. To demonstrate the wide range of evidence relevant to the reincarnation hypothesis I have chosen several well-documented examples of alleged past-life matches from among a larger number of published and private cases. While some of the cases remain anonymous, they all have in common that the evidence presented here has been documented and verified by objective means.

As the experiment progressed, for various reasons, several subjects decided they did not wish to have their names or identifying details published. I have respected their decisions to keep confidential specific details of the research and have changed a few immaterial facts to do so.

Among those who elected to remain anonymous is a possible reincarnation of James Madison (facilitator of the consensus on and ratification of the Constitution and the fourth U.S. President). The project developed scientific evidence on a potential match for Thomas Jefferson (primary author of the Declaration of Independence and the third U.S. President), but the subject does not desire public disclosure at this time. The two people linked with Dolley Madison are given pseudonyms for similar reasons. They are known as Charlotte and Kelly.

The cases below were not chosen as illustrative just because they involve the lives of well-known historical personalities. They are included because the historical record on them is much more extensive than for most dead people. That gives us adequate biographical information to make personality comparisons with their alleged present-day incarnations. No case depends on information from paranormal sources. Most have been initiated by the subjects themselves. In alphabetical order they are:

Gad(Gauguin)/Moshay. Born in California in 1956, Michelle Moshay studied French and travelled to Tahiti and Paris. Not until she was forty-one years old did she realize she had replicated significant features of Mette Gad's life as the wife of Paul Gauguin. She saw her attitudes toward life and children as the logical extension of Mette's experience a century before, including helping a friend relate to Gauguin.

Mette, more than the "wife left behind by Gauguin," established herself as a free and independent woman in Europe during the Victorian Era. She supported herself and her children and became engaged in the world of art business to raise funds for Paul Gauguin. Unintentionally, Michelle took on similar responsibilities. Both women share physical, cognitive, ego, and personality traits with no family-tree connections.

Gauguin/Teekamp. Born almost 102 years apart, Peter Teekamp and Paul Gauguin share more characteristics, as artists and men, than one could expect from "the luck of the draw." Paul was born in 1848 in Paris, moved to Peru as an infant with his mother (his father died at sea en route), and returned to France as a young man. Peter was born in the Netherlands and moved to the United States as a young man. Paul began painting in his twenties and Peter was an artist from his primary school days.

Paul grew up without the presence of a father and Peter's parents abandoned him when he was about seven years of age. Both suffered in search of their artistic calling. Each was successful in business careers, but gave up financial security and family life to fully express their passion for painting. They both tried suicide, their marriages did not survive their careers, and their children were alienated from them.

Gordon/Keene. General John B. Gordon fought the Union Army all over the South during the Civil War. He was with General Robert E. Lee when he surrendered the Army of Northern Virginia at Appomattox Courthouse on 12 April 1865. After the war he got into at the University of Minnesota and then at Harvard (for thirty years). A noted original thinker, he published many substantial books on societal trends before his death in 1968.

Although formally schooled through his Bachelor's degree, Lorin Kee (who lives and works in Tennessee) has also pursued the track of self-education. He studies and writes in Sorokin's style about cultural dynamics, cycles of civilization, history, philosophy, and quantum mechanics. As much of a revolutionary as Sorokin in his day, Lorin challenges fascist tendencies in modern society. While both men are not religious, both love the choral music of the Russian Orthodox Church.

Swidersky/Alexander. Ken Alexander who was born 11 November 1955 in Denver, Colorado is the possible reincarnation of the soul known as Tony Swidersky (born in May 1902 in Pennsylvania). As an adult, Tony worked for the Goodyear Company in Akron, Ohio (the birthplace of Ken's mother) where he was involved in the construction of the USS Airship Akron, on which he later died when it crashed at sea in 1933. He likely knew Ken's grandfather.

At the time of death, Tony was in the Navy. Ken is now a U.S. Naval Captain in the engineering corps. Ken had dreams of the crash of the Akron and scenes in Akron long before he suspected a connection to Tony. He investigated his own case, using public records to confirm details provided by a psychic and from recurring dreams over decades. An evaluation of this case is incorporated into the following chapters.

Stevenson/Tucker Cases. For forty years the Division of Personality Studies at the University of Virginia has conducted the extensive detailed and documented research on the lives of children with memories of previous lives mentioned earlier. Led first by Ian Stevenson and now by Jim B. Tucker, the program has identified more than 2,500 cases which consist of children whose memories, specific knowledge, habits, behaviors, and physical features (to varying degrees) match those of specific individuals who were dead before the children were born.

In the most robust of these cases, the children have had one or more physical traits, including deformities or birthmarks that mimic wounds or deformities in the previous personalities. These children have demonstrated knowledge about the previous life, including family members and many other details that they could not have learned since their birth.

Special Cases. Two other subjects with pseudonyms are Charlotte and Kelly. An extradimensional source, Ahtun Re channeled through Kevin Ryerson, identified Charlotte as the reincarnation of Dolley Madison in 2005. In 2006, Ahtun Re amended that identification in a communication that postulated Charlotte and Kelly were "soul-splits" of Dolley. (See Figure 1.)

While Charlotte had initially expressed skepticism about being the reincarnation of Dolley, and has remained uncertain about the split concept, she agreed to participate personally in the evaluation of her case in this project. Kelly, the spouse of James-II, expressed no interest in the project or being identified as the reincarnation—whole or split—of anyone. Nevertheless, she permitted use of photographs and personal data to illustrate the evaluation process.[2]

Steve Kern presents another intriguing case of alleged soul-splits, involving the deceased singer and musician John Denver. Steve has become known as "Little John" over the decade since John's death. A comparison of Steve's life and personality with the public record of John's life and career provides possible evidence for one or more forms of a soul relationship.

The self-developed case of Barbro Karlen and Anne Frank, published in Barbro's book *And the Wolves Howled,* contains evidence of most of the factors found in the pilot-study cases. Relevant quotes from the book are referenced where appropriate.

James Kent has strong evidence of a past life in the Civil War, but as of now we have not been able to identify a definitive past-life linkage. His case, reported in his book *Past Lives of a Confederate Soldier,* is included to demonstrate good self-directed research procedures and general support for the reincarnation hypothesis.

Another self-identified case for which we do not yet have a definitive past-life match involves Brian O'Leary, astronomer, scientist-astronaut during the NASA Apollo program, and author of *Exploring Inner and Outer Space* and *The Second Coming of Science.* His lucid dreams, "unacquired" knowledge, and body stigmata suggest a case for further research.

A Question of Concept. In the identification of past-life connections in this book, you will not see statements like "Sherrie Laird is the reincarnation of Marilyn Monroe." The focus is on the transcendent nature of the psychoplasm—which may have many lifetimes, of both genders, in different races, cultures, and social positions—instead of a particular personality. When a subject is hypothetically linked to a previous personality he or she is described as the probable present incarnation of the same soul who incarnated in that historical person.

It is more accurate and less pretentious (particularly when historical figures are involved) to say, "The soul who incarnated as James appears to be the soul who later incarnated as James-II." Given our evolving knowledge, the correspondences described in this book should not be taken as proof of a specific past-life identification.

Part I
....
Unsolved Mysteries

How humans acquire new knowledge, develop analytic skills, make decisions, and maintain memories over time remains one of the deepest mysteries of modern science. Most neuroscientists assume the brain is central to this invisible process. They have identified areas of the brain that react to various stimuli, but cannot explain how they transform the bits of data into mental images, habits, and wisdom.

The chapters in Part I review several areas of human experience that are obviously inconsistent with the notion that each new brain learns all that it knows and how to use it as a result of only the child's simple interactions with its environment. Revealing this void in our understanding of how the mind works leaves many unanswered questions.

Where does a prodigy's amazing knowledge originate? How can we explain a child's knowledge or behavior that has not been taught? How can we access memories of knowledge and experiences that are not our own?

Whence do dreams and visions of possible former lives and historical eras come? Why does past-life therapy help people heal chronic psychological and psychosomatic problems? Why do individuals living today look just like unrelated individuals who lived many years before they were born?

After five chapters that elaborate on these and other such questions, Chapter Six reviews surprising correspondences between two lives separated in history. This two-life story sets the stage for a new, personality-based approach to the mystery of reincarnation.

Chapter One

Prodigies and Precocity

In January 2007, Monica Almeida in the *New York Times* wrote about a lecture by Terrence Tao, who had learned to read by age two, taken high-school math classes at seven, college math classes at nine, and earned a Ph.D. by twenty. Now thirty-one, he is among the top mathematicians in the world tackling an unusually broad range of problems, including ones involving prime numbers and the compression of images.

Tao's lecture on prime numbers was broadcast on the Internet and before overflow crowds. Polite and unassuming, he gave a whirlwind tour of the subject which has been a hot topic among mathematicians for 2,000 years. Last summer, he won the Fields Medal, often considered the Nobel Prize of mathematics, and a MacArthur Fellowship worth $250,000. Terrence, like all prodigies, cannot explain why he was born with skills and knowledge no one taught him in this lifetime; he simply knows them.

On 28 November 2004, CBS Television's Scott Pelley compared twelve-year old prodigy Jay Greenberg to Mozart. A composer at New York's renowned Juilliard School, who some say is the greatest talent to come along in 200 years, Jay had already written five full-length symphonies. He says music just fills his head and he has to write it down to get it out. "It's as if the unconscious mind is giving orders at the speed of light." says Jay. "You know, I mean, so I just hear it as if it were a smooth performance of a work that is already written, when it isn't."

"We are talking about a prodigy of the level of the greatest prodigies in history when it comes to composition." says Sam Zyman, a composer and teacher at Juilliard. "I am talking about the likes of Mozart, and Mendelssohn, and Saint-Sans.... He could finish a piano sonata before our eyes in probably twenty-five minutes."

Jay's parents are as surprised as anyone. His father, Robert, is a linguist, and a scholar in Slavic language who lost his sight at age thirty-six to retinitis pigmentosa. His mother, Orna, is an Israeli-born painter. "I think, around two, when he started writing, and actually drawing instruments, we knew that he was fascinated with it," says Orna. "He managed to draw a cello and ask for a cello, and wrote the word *cello*. And I was surprised, because neither of us has had anything to do with string instruments. And I didn't expect him to know what it [a cello] was."

But Jay knew he wanted a cello, so his mother took him to a music store where he was shown a miniature one. "And he just sat there. He... started playing on it," recalls Orna. "And I was like, 'How do you know how to do this?'" By three, Jay was still drawing cellos, but he had turned them into notes on a scale. He was beginning to compose, and his parents watched the notes come faster and faster. He was writing any time, anywhere. In elementary school, his teachers had no idea how to handle a boy whose hero wasn't Batman, but Beethoven.

At the age of ten, Jay was going to Juilliard, among the world's top conservatories of music, on a full scholarship. At age eleven, he was studying music theory with third-year college students. Jay also takes high school courses at another school—courses his parents say he will finish when he's fourteen. Talented composers might write five or six symphonies in a lifetime. Jay has written five by the age of twelve.

Prodigies

The term *prodigy* is commonly used for anyone who masters skills or arts at a very early age. Articles describing the phenomenon often start with phrases like these: "William started learning Hebrew at age three, to master it at seven. He spoke thirteen languages at age thirteen. Giannella, at eight years of age, masterfully conducted the London Philharmonic orchestra. Danielle played piano before she could talk and played Mozart at age four." Some might remind us that Wolfgang Amadeus Mozart himself wrote a sonata at age four and an opera at seven.

Many prodigies have a fragile nature. Sometimes, their knowledge and skills coexist with a mental disability. The book and eponymous movie *A Beautiful Mind* (based on the life of Nobel prize-winning mathematician John Nash) popularized the case of a mathematical genius who also suffered from schizophrenia. He miraculously recovered and achieved a Nobel Prize.

When the disability is extreme, the person may be known as a *savant* or *idiot savant.* The French term savant, like prodigy or *wunderkind* in German, implies someone with specialized cognitive abilities or knowledge which have not been acquired in any normal way.

For instance, Daniel Tammet is an autistic savant who performs mind-boggling mathematical calculations at breakneck speeds. He speaks seven languages and is even devising his own language. The blind American savant Leslie Lemke played Tchaikovsky's Piano Concerto No. 1 after he heard it for the first time, and he never had so much as a piano lesson. According to some estimates, one percent of the general population are savants (as are ten percent of those diagnosed with autism).

Confronted with such stories of an inherent genius in languages, mathematics, or music, one immediately asks the how and why questions. Parents say, "He or she was born with it." Since we know mathematics and music at that level require considerable knowledge and practice, that's the only possible answer. Self-evident from infancy and beyond those possessed by their parents, the skills and understanding cannot be explained by environmental or genetic influences.

In fact, there is no theory that offers a satisfactory explanation. Calling such extraordinary skills "innate faculties" does not explain why they exist. With no conceptual framework into which one can easily place them, scientists resort to labelling them "strange anomalies."

Quick-starts

We often hear of prodigies like those described above because they are so far beyond the norm in a very specialized field or skill. But, how about the similar stories of people in all fields of endeavor who start exhibiting extraordinary ideas, skills, or interests at a very young age? All that separates these "quick-starts" from prodigies is a matter of degree.

When Barbro Karlen's book of childhood poems was published at age twelve and became wildly successful, she was hailed as a child prodigy by the Swedish media. From her perspective, she wanted to be a child like everyone else. For her, powerful writing came as natural as riding horses that won national dressage competitions.

Tennessee musician and writer Dalton Roberts wrote the author, "At an early age I discovered I had a natural ability to write poems and songs.... (They) usually come to me with words and melody fitted together. I may do a Tom Horner-style tweaking, but ninety percent of them end up just like they were born. I can claim no credit for this ability. It has always been as natural as breathing.... I think it must surely come from another life I have lived."

Willie Nelson who wrote his first song about cheating hearts at age seven, well before he had experienced romantic love, offers his personal explanation for this apparent musical birthright. On public radio he has reiterated his belief that he was reincarnated with this reservoir of music. Are Dalton and Willie's musical skills any different from the three-year old's who declares she is a dancer and proceeds to demonstrate the skill to do pointe before her muscles are supposed to be capable?

Perhaps our category of "quick-starts" should include muscular skill as well. In the case of an Indian girl named Swarnlata described by Jim Tucker, her repertory of "past-life knowledge" was not limited to foreign Bengali songs. Her untrained body could perform the intricate dance movements associated with the music.[1]

Young children who also report memories of previous lives may indicate they came into this life to achieve a specific developmental objective (and from birth set about to do it). This is sometimes illustrated in the persistence of early play. Tucker described the play of a young boy who reported having been a baker in a previous life and seemed inclined to become one in this life.[2]

Precocity

A friend has a large cardboard box of realistic sketches he made of animals, flowers, and buildings before he started school. He says he always knew how to draw and that his hand didn't need much practice before it could reproduce what he saw in his head. Peter Teekamp, placed in an orphanage at age five made himself welcome among his classmates by drawing their portraits. I recall my own dramatic reading skills that appeared when I entered school from a home with no books and parents who had no time to teach their children to read.

A young girl from a rural, introverted family handles herself with aplomb when in a competitive public arena. A small boy reacts with compassion towards the suffering of a strange family. How did they learn to do that without prior experience? The most frequent answer is: "They are precocious." While this provides a label, it does not offer an explanation.

Some refer to children like those described here as "old souls" or "indigoes" to suggest that they are more mature, especially in psychological or moral development, than their peers. Such children react to events in an unexpected manner given the age and experiences we know they have had so far in this lifetime. There is no evidence that these children (or souls) represent anything more than the normal percentage of people who score high on any number of tests. Perhaps we should begin to think of all children as precocious, differing only in degree.

Unexpected Attitudes. An *Atlantic Monthly* article on beliefs about the survival of consciousness described one area where children from birth have very definite beliefs. Early in life, before they have been exposed to such information, some children report a belief in the continuation of a form of existence after the physical destruction of an animal or person's physical body. The article suggests they had not been in a situation to learn this belief before they expressed it.[3]

Yale psychologists Paul Bloom (author of the *Atlantic* article) and Deena Skolnick Weisberg described this as an "intuitive" childhood belief in the immaterial nature of the soul. They also learned that the children resist adult arguments that contravene this intuitive sense of reality, which survives to adulthood.[4]

Tucker's research at the University of Virginia corroborates the above findings with case studies of young children who reassure others that the deceased person is no longer in the grave. That they "know" it is reflected in their refusal to be afraid or sad.[5]

While these various researchers had different objectives, their common finding is that children appear to have an innate predisposition towards a multidimensional view of reality. This is consistent with our view that such attitudes precede birth.

In Search of Explanation. While only certain cases of child prodigies get media attention, teachers tell us they can be found in classrooms around the world. The most publicized are in narrow fields of endeavor, like the above-mentioned musical virtuosos or mathematical geniuses. However, extraordinary behaviors in many children, and unusual ones in others that seem close to the societal norm, are equally mysterious in one respect: These skills are exhibited by children whose genes, environments, and experiences, cannot be shown to account for them.

In this chapter we have looked at some of the most self-evident cases of individuals whose birthright was clear very early in childhood. While neuroscientists can describe the brain activity that accompanies amazing feats and psychologists can define the various steps of the mental processes, they cannot explain their origins.

If we look systematically at all the untutored and inexplicable behaviors of most children, we begin to see that all of us are prodigies in less obvious ways. We find child virtuosos in computers, gardening, swimming, human relations, tinkering, dancing, talking, drawing, etc. All children seem to have some knowledge and skills well before she or he could have learned them as a result of external influences.

Chapter Two

Anomalous Knowledge

Perusing popular books on reincarnation, one finds numerous examples of people with specific knowledge that they had no apparent opportunity to learn in this lifetime. Books by best-selling author Joan Grant contain intuitive details about life in Greece (2000 BP) and Egypt (4000 BP) that were later verified by scholars. A Percy Shelley scholar provided an example: When the poet was walking with friends in an area for the first time, he said, "Over that hill, there is a windmill." When they got to the other side, they saw a windmill. An American visitor to a European castle told me of a time when he spontaneously said to the guide, "There used to be a door here." The guide confirmed to him that historical records showed that to have been the case.

While other hypotheses are proffered to explain such anecdotes, they seem analogous to reincarnation stories like the following: Blanche, a very young British girl, was taught a French song by her nanny shortly before she died. Three years later Blanche's mother, again pregnant, gave birth to a daughter. She, at age six, sang the same song without ever having been taught it.

A friend wrote me, "When my daughter was five, she told me a long story that included, 'The wagon turned over. Going down the steep bank. And the horses died.' I told her, 'I don't think that ever happened.' and then she said, 'Yes it did! You were there!' A moment later she said, 'Oh, I remember. That was when I was the other girl. You weren't my Mommy then.'"

Such memories are not always specific enough to point to a previous lifetime, but they are so ubiquitous that many are convinced that they must result from a form of reincarnation. Some are geographically specific enough that they point to reincarnations within the same family or neighborhood. An example came from a friend and colleague, Malou Zeitlin.

Talking to a New York-state neighbor in 2006, Malou learned how he as a little kid had "spooked" his father. The young man's family had lived in the area for generations as farmers. On a walk one day, he asked, "Dad, where are the two houses that used to be here? One had a porch that was coming down. What about the road, where did it go?" He exclaimed, "There WAS a road right here, we used it with the wagons!"

According to this young man, who evidenced neither knowledge of nor interest in reincarnation, what "spooked" his dad was that there used to be two houses right where he said they were. Further, one of them had a porch that was falling down. An old wagon road had been where the kid said it was, but had reverted to brush. His dad kept repeating, "No way you could know that, you weren't even born yet!" The cases in this book include many such examples.

Finding Specific Places. The Swedish family of ten-year old Barbro Karlen traveled to Amsterdam for the first time while making a tour of Europe. They decided to visit the Anne Frank house. When her father picked up the hotel telephone to call for a taxi, Barbro told him they didn't need a taxi as she knew how to get there. In a ten-minute walk through the winding streets of Amsterdam she led them to the house. She then revealed an intimate knowledge of the house, including a room where the Franks had hidden from the Gestapo and what had been in it.

George and Darlene Mettler first visited England in 1972. Wandering without a map, they found themselves on the Strand near the Duke of Somerset's sixteenth-century palace. English scholar Darlene suggested a visit to the Dr. Samuel Johnson House, where he wrote his *Dictionary of the English Language.* When George agreed, she said, "Let's get a map with the directions." George said, "Oh, I know where that is."

Without knowing why or how, he led them up the Strand to Fleet Street where he turned left onto New Fetter Lane. From there, he unerringly navigated the maze of alleys to 17 Gough Square where Johnson had lived from 1748 until 1759. (The couple later reported they had "felt at home" from the first day in London. A few years later, they responded to a strong pull to pick up and move from Florida to Wimbledon to pursue their respective writing and scholarly careers.)

Wayne Peterson, who had developed a strong sense of a past-life connection with Francesco Foscari (1373–1457) who was a doge (leader) of the fifteenth-century Venice Republic, was on a trip to Italy. While interested in learning more about Foscari, he did not have many details to work with. In Venice, he walked up to the Foscari home and Francesco's grave site without having a map or directions to it.[1]

Someone Else's Name. In researching unaccountable utterances, one frequently finds examples of children calling themselves by a name unknown to their parents. For instance, Barbro Karlen told her parents from the beginning that her name was not Barbro, but "Anna Franke." Peterson has reported that as a three-year old he would not respond to the name Wayne and told his parents, "My name is Francesco Foscari."

As a young child Peter Teekamp spontaneously kept repeating a name like "go-gone" in his Dutch accent. His parents had no idea what it meant, but he refused to drop it, shouting it at age ten and pestering his teachers about its meaning at age fifteen. As an adult, while considering a Paul Gauguin connection, he realized the French pronunciation of Gauguin sounded very similar to his childhood Dutch watchword.

How do we explain the immediate deep connections that occur for no apparent reason upon first hearing a name or seeing a portrait or photograph of a deceased person? During my research, I discovered it was a widely reported experience. The same sort of emotional reaction occurs when many people meet someone they have never seen before, yet immediately feel they know them. Often people talk about having walked around a corner in a strange place and recognized parts of the scene before them. Psychologists offer various explanations for this type of spontaneous "knowing," Could it be evidence of memories from previous lifetimes?

Language. One of the most widespread mysteries related to reincarnation involves the spontaneous knowledge of foreign languages. In these cases, the small child typically, even before mastering his mother tongue, demonstrates skill in a foreign language. Such languages, when unknown to the local community, may have contributed to the etymology of the "gift of tongues" connotation of the Latin-derived term *glossolalia.*

Ian Stevenson's vast collection of data includes the Indian case of Swarnlata. At age five or six she was performing songs and dances no one in her family or neighborhood recognized. She retained these memories until well into her twenties. As she matured, she and others learned they were Bengali (from another part of India far from her birth family's culture). Also, in a Lebanese case, Imad was precocious in French—the only French speaker in his family—as was Ibrahim, his reputed previous personality, in the earlier life.

These examples of "unacquired" or precociously learned languages appear prior to the person having training that would account for them. Interestingly, academic research totally unrelated to reincarnation suggests that all learning of language may be based on prior knowledge. Decades ago linguist Noam Chomsky confronted evidence that children seemed to master any language faster than they are taught. He believed only a pre-existing capacity could account for it.

Subsequently, many linguists have empirically reached similar conclusions: (1) A "child learns language with limited stimuli," known as the poverty of evidence. (2) Learning input during the period of early language acquisition is circumscribed and faulty. (3) The quality and quantity of input is less than the output. A person knows vocabulary and language rules without instruction or direct evidence of learning (labelled as "knowledge without ground"). (4) They conclude, "The character of the acquired knowledge may be largely predetermined."[2]

These four examples from research literature in linguistics sound much like the ways reincarnation researchers describe the cases of spontaneous or precocious acquisition of foreign languages. The linguists do not offer an explanatory theory that accounts for their findings. Could some of what the child knows in spite of "limited stimuli," "circumscribed acquisition of knowledge," "more output than input," and "knowledge without instruction or direct experience" come from a pre-existing reservoir of knowledge and skills brought forward by reincarnation?

While these examples describe children who develop foreign languages at an early age, a possibly related phenomenon appears in adulthood. Known as *xenoglossy*, it involves someone, usually in a trance state, who begins speaking a language totally unknown to the speaker. One example in the experiment is that of Kim Adams, who in his forties, began to speak a tongue that neither he nor others around him recognized. He does not enter a trance state, and is both conscious and responsive while speaking the language.

It sounded like an Early American language. He was later able to learn it was a real language spoken by the Latgawa people known formally as the Confederated Tribes Rogue Table Rock and Associated Tribes. Kim met a living speaker of the language, John Grey Eagle Newkirk, the chief of the tribe, who understands him.

When the language first erupted and no one understood it, Kim's partner Dee Loecher did not know how it happened, but offered her interpretation of it. Later John Grey Eagle confirmed the general accuracy of her translations. Kim and Dee have now come to understand that Kim is serving as the voice of a nineteenth-century incarnation of Kim's soul as a Latgawa Indian. They believe Dee is the reincarnation of that deceased chief's wife who was known as Shining Moon.[3]

Unexpected Choices. Sometimes apparently inherited knowledge is just below the surface of awareness and gets expressed in actions rather than words. In the 1970s Michelle Moshay had no idea why she resisted schoolmates' pressure to take Spanish classes in Southern California to study the French she preferred. She does not know the source of her confidence to teach herself how to play the piano as a young girl. Later, learning of a possible link with Mette Gauguin, she saw her choices may have had antecedents in an earlier lifetime. Mette learned to speak French in the 1870s and taught it and piano to financially support her children.

When Michelle was a young girl, her early images of being a mother scared her. Instinctively (her term) she anticipated problems with supporting children. Even with no experience of such a situation in her mother's past, she feared she would lose a child to an early death. Decades later Mette's story of hardship rearing five children alone in the nineteenth century gave meaning to her fears.

Peter Teekamp had no idea where his childhood notion that he would go to America came from. He says it was always there even though he knew no Americans as a child. Even later when he thought about it, he would conclude that it was an impossibility for an orphaned teenager with no connections and means to get to America. After he married an American woman and settled in the United States, he learned of Gauguin's belief that he would have to get to America to be accepted for his art. (See Figure 2.)

Members of a Soul Family?

Profiles of a young Dolley and a young Charlotte. Charlotte's cosmetic nose lift affects phenotype.

Images of Dolley and Kelly as young mothers. Hair styling and dental care affect phenotype.

Fig. 1

Four Individuals
Two in 1897 and Two in 1997

Paul Gauguin   Peter Teekamp   Mette Gauguin   Michelle Moshay

Fig. 2

Of the subjects in this study, Peter describes his past-life legacy the most succinctly: "My present life reflects the past. The roots of my daily life today are in a life lived before." He goes on to say that if we don't recognize that and learn from our past mistakes we don't deserve the future.

Jeff Keene, the apparent incarnation of Civil War general John Gordon, found over the course of several years that he had fragments of "inherited knowledge" —facts that could not have been learned in this lifetime. For instance, not a rider himself, he was out with friends who decided to rent some horses. When the trainer asked the group for its best rider, he impulsively raised his hand and said, "Yes, I'm good."

As a result he was given Rebel, the most animated steed in the group. To Jeff's surprise, when he mounted the horse, he knew exactly how to control him.[4] Years later Jeff learned of his past-life tie to Gordon and Gordon's life as a cavalry officer.

On another occasion, during a visit to the present-day replica of the Appomattox Court House, Keene saw a poster illustrating the 12 April 1865 surrender of the Army of Northern Virginia. He immediately "knew" that a battle flag shown being held by one of the troops was not at the actual ceremony. When he questioned it, a park ranger confirmed that, although the flag had already been adopted as the official National Confederate Flag, it had not been issued for use in the field at that time. Therefore, it could not have been at the ceremony; Keene was correct.[5]

In the beginning of another Civil War case, James Kent realized he knew how to place explosives on bridge structures to bring them down. As a U.S. Postal Service worker in this lifetime, he had neither access to nor experience with explosives. During subsequent research on Civil War ordnance and battle tactics, he learned that his "imagined" knowledge was valid.

Cryptomnesia, a concept favored by many skeptics who challenge these cases of apparently unlearned knowledge, implies one who believes he is having an original thought or memory for the first time has in fact been exposed to the material and forgotten it. In the cases discussed here I have tried to preclude that possibility through establishing that the first expression of the specific knowledge preceded the subject's opportunity to learn it.

Chapter Three

Dreams and Aspirations

Unbidden images universally pop into the human mind as people go about their daily routine and as they sleep. They come as a dream, vision, deja vu, fantasy, aspiration, message, or second sight. Regardless of the label, they are a ubiquitous, unexplained human experience. Some individuals consider them to be the result of random firings by the brain's neurons getting rid of overloads. Other people consider them to be special messages from a divine realm. Most of us, however, have no idea where they come from. But, many of us think they have meaning.

Could some of these images and the information contained in them be connected to people and events in the past? The claims that many dreams and aspirations are based in previous lives make them part of this experiment's effort to solve the mystery of reincarnation stories.

Dreams

Ken Alexander is an interesting example in this regard. He claims to have a very intensive memory of his dreams going back to when he was three years old. He had no idea of what to think about these vivid images, some recurring, until the year 2000 when he read Michael Newton's book *Journey of Souls*. After finishing it, he states that he, "reluctantly accepted (the possibility of) reincarnation," although, "frankly, the dreams made no sense and did not provide a clue as to the overall substance and sequence of lifetimes. I tried regression and it was not that useful."

After the events of "9/11," one of which retroactively seemed to have foreshadowed the attack on the Pentagon, Ken reports he "attempted to write down as many of the dreams suggestive of past lives as (he) could in a series of letters to a friend." Keep in mind this happened in 2001, before he obtained clues from a 2005 session with trance channel Kevin Ryerson (mentioned later) enabling him to piece the Swidersky story together.

Two of those dreams involved his death. One that occurred on 4 April 1981, when he was in naval-engineering officer training, would subsequently convince him of the veracity of a connection to the life of Tony Swidersky. He later realized the dream had occurred exactly forty-eight years after the crash of the U.S. Navy Airship USS Akron on 4 April 1933. Subsequently he learned his dream details reflected actual events during the last few minutes of the Akron's life. (See Figure 3.)

Dream Connects Two Real Lifetimes!

Ken Alexander
U.S. Navy - 2004

Tony Swidersky
U.S. Navy - 1932/33

Fig. 3

The death dream had him in a German-American automobile, starting an uphill, westward drive at dusk. There was no road, but he knew where to drive and could see sunlight coming up through cracks ahead of him. He was very tired, so he stopped at a friend's house and asked if he could sleep in his bed. He was pulled out of bed as soon as he lay down and found himself back in the Opel. The steering wheel and brakes did not work and the car rolled backwards very fast.

When the car in the dream stopped, it was hanging over a cliff at an angle. After dangling for a moment, the Opel fell off the cliff (going backward) and he felt himself alone. When he hit the water it seemed in his dream that the sun came out. He saw dozens of other people struggling in the water, but knew he was dying.

Later research revealed the German/American-built Akron lifted off the ground at 7:30 p.m. on 3 April 1933 and eventually sank just after midnight on 4 April. It headed west (like the Opel) from Lakehurst, NJ towards Philadelphia. Swidersky was the helmsman at the wheel. When he got tired, he slept briefly in a crewmate's bunk as was the custom in the cramped cabin (as Ken did in his dream). When a storm hit at 11:00 p.m., the ship had difficulty maintaining altitude, so he was called back to the wheel (having worked on its design and manufacture). It crashed into the Atlantic Ocean tail first (as the car in Ken's dream), assuming an angle of 20–45 degrees before sinking into the ocean. The crew found themselves in the water.

Accepting that Ken's dream mimicked reality, you might say it resulted from his having read and forgotten published articles about the Akron. He asserts he had never heard of the Akron until the 2005 channelling of Ahtun Re. You might still protest that he was simply picking up the long-lost memory fragments energized by the dying crew members. However, recovering someone else's memory fragments does not account for all the other statistically unlikely correspondences between Ken and Tony that are described in this book.

The first details about a possible past-life for James H. Kent surfaced at age thirty-nine when he began a series of dreams in which he fought as a Confederate soldier. A platoon sergeant in a Virginia regiment, he dreamed of being wounded during a battle in Maryland. Over several years of dreams, during which he also tried past-life regression with no success, he could not obtain information to identify the sergeant. However, his dream memories turned out to reflect the life of a real nineteenth-century person.

Prior to his dreams Kent had only a superficial knowledge of the Civil War and little interest in the subject. When he decided to try to independently verify if his dreams were from the past, not something that he might have seen or read, he refrained from reading related materials. It felt strange that in one dream he and his men had no uniforms, but later historical research proved to him that this was the case during parts of the war. In another dream, his image of a young

Confederate officer wearing a tan uniform with a red sash (instead of the grey uniform Kent the dreamer had on) later found confirmation in a Civil War book of uniform illustrations.

When he dreamed that a Federal soldier who could have easily killed him in a skirmish deliberately shot over his head, he assumed he had made it up. Later he found historical reports revealing this was a frequent practice on both sides in the war. His dream of a small drummer boy's behavior and leggings was also later authenticated by historical research. Details of a unique battle flag he saw in a hypnosis session were also later confirmed through Civil War sketches.

Subsequent research also confirmed for him, much to his surprise, that a Virginia regiment like the one he dreamed about had fought in Maryland. But the biggest surprise to him was that the Confederate general he saw badly wounded on the battlefield and called out to by the name "Sam" turned out to be a real person. The life, campaigns, and battle wounds of General Samuel McGowan are well-documented in the National Archives. Kent's dreams of hand-to-hand combat and the carnage of the skirmishes fought that day parallel historical events.

Most dreams and visions are less precise, including most of those that arise in hypnotic sessions. Nevertheless, they leave the person with lingering questions about their meaning. In a 1982 hypnotic-regression session, Brian O'Leary had a vision of himself in a jail cell being given a death summons. He had never before experienced such visions. But, he interpreted it to be in the Vatican and related to something he had done that the Church did not like. He then saw himself, as if out of his body, tied to a stake and being stoned or burned in a courtyard.[1]

Eleven years later, rushing to see the Sistine Chapel in the Vatican Museum, his eyes were drawn out a window to a courtyard. To him, it looked exactly like the one he saw in his hypnotic vision. He says, "shivers went through my spine and I felt mildly nauseous."[2] This confirmation led him to become more interested in the possibility of past-life implications of his visions.

Aspirations

Many children express definite ideas of a "life goal or purpose" at a very young age. Their visions seem to be internally generated, generally unrelated to the experiences and expectations of the family and their immediate culture. The origin of such ideas remains a mystery. Equally mysterious is why so many of these childhood fantasies and games materialize later in life?

Famed neurosurgeon and inventor of medical technologies Norman Shealy, at age four, declared to his parents that he was going to be a physician. Susan Kolb, one of the top plastic surgeons in the U.S., can offer no explanation for her childhood assertions to anyone who would listen that she was meant to be a plastic surgeon. She describes it as something like a voice, a part of herself speaking to herself.

A genuine belief in this voice, whether it was real or not, could motivate her to enroll in the proper course of medical studies. But, how can one attribute to that "voice" her ability to graduate in the top 1% of her class in one of the nation's most demanding medical schools? And, to be certified at the youngest age on record by the American Board of Plastic Surgery?

Can a simple belief in one's aspirations make a world class neurosurgeon out of four-year-old Shealy and a topflight plastic surgeon out of the child that was Kolb ? It appears that some part of their child-minds already had accurate information about the potential of their own talents.

This seems to have been the case with former scientist-astronaut Brian O'Leary. When in grade school in the 1940s, he declared to his family, teachers, and classmates his interest in space travel, going to the moon and other planets. The editors of his high school yearbook labeled his class picture with "He's going to the moon." Brian went on to earn a Ph.D. in planetary astronomy and was accepted by NASA as a candidate for the Apollo Mars mission. When that program was cancelled, he joined Carl Sagan at Cornell to work on unmanned planetary programs.

Could these precocious aspirations that reflect actual capabilities be grounded in an innate awareness of prior experience that qualifies them to proclaim a realizable future? Are they any different from other early-childhood activities? Do patterns from a previous life express themselves through playing games? Some interesting examples suggest that may be the case.

As a young boy, twentieth-century fire chief Keene found himself building play forts for mock battles with friends, like untold numbers of kids across America. The only difference was that instead of simply digging a hole and piling the dirt up around it as barricades, he scrounged scraps of wood to cover his structure. In his forties he learned of similar structures, called "bombproofs," from studying Civil War history.

In the Civil War, Confederate officer Gordon had a lot of experience with such battlefield structures. They were constructed by holes being dug into the ground and roofed over by whatever wooden timbers or planks were available. These shelters provided varied degrees of protection from the artillery shells fired into the embankments by the opposing side. Many soldiers later credited a bombproof with saving their lives.

Stevenson reported in his *Twenty Cases* book the story of Jasbir in India who believed he had a previous life as a Brahmin (Hindu high caste). As a child, although he was born into a lower caste, Jasbir wore a string around his neck in the distinctive habit of a Brahmin. In the case of Imad from Lebanon, Stevenson reported he had shown an early interest in playing with guns and hunting. Ibrahim, Imad's alleged previous incarnation, had been a hunter and kept a rifle and shotgun in the house.

Tucker's survey of these and similar cases for his book *Life Before Life* included the case of Parmod who played at being a biscuit shopkeeper. Research revealed that Parmod's other-lifetime memories were associated with a biscuit shopkeeper. Such games relate to life interests.

Parents, psychologists, and educators often wonder why some people seem to have been born with an unexplainable ambition in a direction not seen in other members of the same family. Scientists cannot account for these childhood differences in motivation related to their chosen areas of creativity. We can resort to phrases like " the traits are a product of evolution," but we have no evidence that they result from random, but selective genetic mutations. Being highly motivated by a combination of specific goals and having the energy to accomplish them doesn't lend itself to a neat "Gene X2Y9 = Behavior" equation.

The wide array of self-identified aspirations that come to fruition cannot be accounted for by the variables of genes, gender, family, and culture. People like William Jefferson Clinton, Oprah Winfrey, Martha Stewart, Tiger Woods, and Condoleezza Rice remind us that individuals start life with a set of unique, internal motivations that do not depend on external stimulation or social pressures. Could they, and the rest of us living today, have started with a set of predispostions already well established before birth?

Chapter Four

Past-Life Healing

While searching for evidence that laymen consider supportive of a theory of reincarnation, I received an e-mail from Canadian Jonathan Kolber. He wrote, "I had harbored hostile attitudes towards France, a country with which I had no apparent connection whatsoever. I regarded the French as irrational and prone to mob rule. I also had a phobia of crowds."

During a spontaneous "memory" of scenes of what he thought occurred in eighteenth-century France, Kolber reported he had heard a waltz and saw the orchestra playing it. Later, he was able to sing the waltz tune well enough that it was recorded on a cassette and given to a professor of classical music. The professor declared it to be the work of "an unknown composer," probably circa 1800. Kolber claimed he had never composed music, much less a waltz, and had not even taken a music class after grade school.

After this seeming "remembrance" of a life as an insignificant member of French nobility, whom he believed to have been born in 1756, Kolber reported his childhood, irrationally hostile attitudes towards France disappeared. For the first time in his life, he said, he was able to enjoy vacationing in Quebec and listening to the French language.[1]

Were the images simply a hallucination? Was his musical cassette mistaken for a real composition? Perhaps. But, the subsequent personality changes were real. While a case like this doesn't prove the person's reincarnation, it presents science a challenge. Could his energetic or emotional connection to a scene in eighteenth-century France, that released hostile attitudes and a serious phobia, be as real as one's memory of a childhood trauma?

Because memories like these are difficult, if not impossible, to authenticate, it does not necessarily mean they are false. That we cannot use them as direct evidence to support a specific reincarnation case does not prove such reports are worthless. In fact, a therapeutic industry now functions based on belief in the veracity and power of recalling past-life memories. To understand their potential value to the reincarnation hypothesis in this book, let's go back in historical time.

Cathartic Release

Vienna, Austria in the 1890s found Josef Breuer and Sigmund Freud pioneering the psychological healing approach known as psychoanalysis. Breuer had developed a cathartic process that helped patients gain relief from a debilitating psychological complex. He and Freud learned that symptoms could be reduced by exposing unrecognized emotional links between a repressed memory of an actual past event and its current psychological and physical effects.

Freud developed the concept of free association as a technique to recover the long-lost feelings associated with a traumatic incident. He used hypnosis to get past the patient's resistance to recalling what had been such a shock in the first place. He would then take people back to earlier events, particularly things that happened in childhood. He, Carl Jung, and Alfred Adler found the process worked as expected.

These techniques have been modified by psychiatrists and psychologists over the past century, but in its many forms the essential discovery remains validated. The suppression of memories of traumatic events adversely affects one physical and mental health until the blocked energies are released through conscious resolution. If the recovery of suppressed childhood memories helps people solve psychological issues in this lifetime, should not recovery of possible memories from an earlier lifetime have the same effect?

Past-Life Therapy

In the early 1970s, Roger Woolger, a graduate of Oxford University and a certified Jungian analyst, critically reviewed the work of others on reincarnation. Later in the decade, still skeptical, he agreed to experiment with a colleague's technique for regressing oneself to a possible past life. Roger found himself experiencing unexpected images and feelings.

Sometime later, in his practice of regressing clients to early childhood, he found that some slipped into memories that seemed to come from earlier eras. Those memories, once awakened, seemed to allow the patients to cope with their shadow sides of unsociable, violent, angry, or brutal tendencies. A time of "reliving" memories of traumatic events in a postulated previous life obviously facilitated the release of psychological burdens in this life.

When I talked to Roger more than twenty years later he rermained skeptical about reincarnation theories, but remained convinced of the psychological healing power of recovering alleged past-life memories. His book *Other Lives, Other Selves* describes cases that have become classic examples of how clients have gained psychological relief through what they believe to be the remembering and re-experiencing of the events from past lives.

One woman could not bear to leave her cats unattended until she "recovered" a memory of a previous life in which she had accidentally killed her child by neglect. Another, a successful painter, suffered from immobilizing guilt about a personal move that left her mother behind. She "recalled" that, as an unsuccessful painter in a former life, she had committed suicide after letting a child die while under pressure to produce a painting. The recall allowed her to get back to work.

The disappearance of physical symptoms is also frequently associated with reliving trauma in past lives. Roger recounts the stories of a man with a hysterically paralyzed arm, a woman with a serious illness affecting her back, and an osteopath with incurable sinusitis. In each case, the alleged remembering of specific lives enabled physical as well as emotional healing. One learned of a past life involving an accident resulting in a broken arm. Another recalled a past life of a pioneer woman who broke her back in a wagon wreck. The osteopath became well after resurfacing the feelings he had suffering from a cold at camp while his mother lay dying at home.[2]

In the last quarter century, similar cases have been resolved in the offices of many therapists, including well-known practitioners like Bruce Goldberg, D.D.S.; Edith Fiore, Ph.D.; Brian Weiss, M.D.; and Adrian Finkelstein, M.D. Critics charge that the search for past lives with hypnosis or other guided techniques may simply create a self-fulfilling prophecy. If that is the case, the therapist and client unconsciously conspire to imagine roots in the past to justify the change that the client and therapist know he needs to make.

Regardless of whether the alleged memory recovery is imaginary or authentic, one cannot deny the beneficial effects on all these lives. Hysterical physical symptoms, depression, phobias, and other psychological problems quickly and painlessly dissipate when the so-called, past-life information is revealed.

The insights of Carol Bowman's children on a 1994 *Oprah* show described in her book *Children's Past Lives: How Past Life Memories Affect Your Child* represent the essence of this past-life healing phenomenon. Her daughter Sarah challenged a skeptical psychologist on the television studio set as follows:

"I say it doesn't matter what it is. Maybe it's not religious... whatever you said that was. What matters is that I was afraid of fires and I'm not anymore. And Chase had problems with his wrist and he had a fear and he's over that. What matters is that it helped us."[4]

Past-life-healing reports like those described above may offer clues to a particular previous lifetime. Together with the additional types of evidence described in subsequent chapters, these clues can help produce tangible evidence of reincarnation in general and a specific past-life match.

That the past-life regression technique produces psychological changes in the individual is consistent with and supports the complete personality reincarnation hypothesized in this book. If the effects of powerful emotional events experienced in a previous life are still discernible in this life, they provide concrete support for the notion of inherited personality predispositions.

Chapter Five

Dead Doppelgangers

Many of us have seen or heard others claim to have seen someone who looked exactly like us. In one such example, a dear friend who had known me for years excitedly called from New York City. She recounted the experience of just "having seen" me on the street—a man who looked so much like me that she "knew" it was I, until she had grabbed his arm and looked into his eyes. She told me this doppelganger of Paul Von Ward had all my physical and facial characteristics except that he was cross-eyed. When she realized it was not me and told him she was sorry to have mistaken him for a friend, he graciously assured her it had caused her no problem. She said even his polite manner mirrored my own.

Please take a moment to examine the pairs of photographs in this book. They compare the portraits of people in the past with people today to whom they are not genetically related. Like doppelgangers across time, this photographic evidence of physical correspondences between subjects in this project and their hypothesized earlier lives seems statistically improbable. There must be a natural and plausible explanation for this unresolved conundrum.

Ian Stevenson began his research on children who had memories of the lives lived by people who had died before they were born and who lived in a different part of the country or world from their families. From 1961 until he died in 2007, he and colleagues accumulated over 2,500 cases of children who reported memories of information, places, and events associated with earlier lives and times where the specific details were later confirmed to be accurate.

Several years into his research he became aware that many of these children had birthmarks and birth defects that corresponded to marks, deformities, or wounds on the bodies of the deceased individuals. In later years he also noted similar physical characteristics such as skin pigmentation, facial features, physiques, and racial characteristics. His book *Where Reincarnation and Biology Intersect* provides the reader with detailed cases.

Other researchers mentioned later on have added cases with uncanny physical similarities between current subjects and their alleged previous incarnations. This chapter describes some of these strange and unexplained physical matches for which science has no ready answer. Part II describes possible explanations for these correspondences and posits a plausible hypothesis.

Among the Stevenson cases, one young boy alleged he had been killed in a previous life by a blow to the back of his skull. He was born with an indentation at the base of his skull with a birthmark in the same location. Subsequent research identified the previous personality and discovered he had in fact been killed as the child described. In another case, a young child claimed to have been shot to death in his previous life. His body had a small birthmark at each point where a bullet entered and exited the body of the deceased individual.[1]

Such marks are often not just skin discolorations, but irregular raised patches of skin.

Physical features carried forward also include missing or malformed limbs, fingers, or toes. Among Indian cases reviewed by Tucker, one-third had such past-life related birthmarks. Eighteen percent were corroborated through medical records for the previous personalities.

Tucker further reviewed cases from Asia and Alaska where people sometimes either predict rebirths that include identifiable markings or actually mark the dead body to see if the predicted new-born carries them.[2] These cases raise the question, "What sort of mechanism could account for these non-random similarities?"

If not born with such marks, the body sometimes seems to conspire with events to acquire them later in life. This appears to have happened in Keene's life. His right leg developed spider vein clusters in two places where Gordon had been wounded. Keene's left forearm has a surgical scar at the point of a wound on Gordon's arm.

Late in his career, Stevenson wrote, "I have become convinced... that is some cases unusual facial features of a subject correspond to similar features in the face of the person whose life the subject claimed to remember."[3] He focused primarily on unusual features.

However, he also noted that a subject who remembers the life of a person from a different race may resemble that race more than the race of his biological family. He suggested further study through the use of photographs.

Other researchers have followed up on this suggestion using photographs or portraits from historical personalities. In his research, Semkiw made a subjective, facial correspondence of such images as one of the criteria for assigning a "confirmation" to his past-life identifications. One can see an array of his proposed matches in the book *Return of the Revolutionaries*. Finkelstein used facial comparisons to great advantage in his presentation of the Laird/Monroe case in his book *Marilyn Monroe Returns*.

In the self-researched case of Jeffrey Keene, he noted early on the high degree of correspondence between his own face and photographs of John Gordon. When he accidentally found a picture of Gordon in an issue of the *Civil War Quarterly*, he recognized the face as one he knew well. He wrote, "I shave it every morning."[4] Anyone can see the same jutting jaw in both men's profile. A closer inspection reveals two mouths shaped like a horizontal Cupid's bow.

Consistent with Stevenson's research in Asia and the Middle East on the appearance of marks that reflect the exact areas of wounds made on the body in a previous life, Keene's face is a virtual map of Gordon's battle scars. He has markings on the right and left side of his face that correspond to the entry and exit wounds on Gordon's.

The triptych at Figure 4 uses portraits of James Madison and James-II around the age of thirty-two. The degree of the match is even more startling when one considers that both are portraits painted by artists separated by continents and almost two centuries in time. The older was done in 1783 by Charles Wilson Peale in Philadelphia for James's second serious sweetheart and the newer was sketched in 1972 in Paris by Place du Tertre artist Reyes for James-II's second wife. In 1972, James-II had no personal interest in reincarnation or past-life connections.

The similarities do not stop with the face, but include the body as well. In the case of James and James-II, surviving portraits and third-party reports make physical comparisons possible. Both were below average for their times, with slight (ectomorph) bodies. Friends put James height at between 5'2" and 5'6." He was reported not to have weighed much above 100 pounds. James-II reached adulthood weighing less than 120 pounds and just reaching 5'6."

Two males, same age, separated by two centuries. One psycho-energetic genotype?

James - 1783   James-II - 1972   One genotype?

Fig. 4

Michelle and Mette were both about 5'6" and inherited mesomorph body types. Mette was described as a "mannish, sturdy woman" and Michelle describes herself as always having "a weight issue." Peter Teekamp and Paul Gauguin were about the same height and weight, each with ectomorph body types. (See the photograph of both couples at Figure 5.)

Semkiw has published photographs of possible present and past-life matches who apparently changed genders. In them, we find the same underlying features manifesting in somewhat more feminine or masculine bodies, depending on the direction of the switch. In support of this thesis, Stevenson noted distinctive leg shapes appeared to carry forward even when the subject's sex was different from the recalled lifetime.[5]

These few examples, to be extended in coming chapters, suggest that if we could collapse two moments in time and put our subjects and their hypothesized earlier lives side by side they would look like twin actors in two different plays. They would in many instances look more like time-travel doppelgangers than resembling their own parents or siblings.

Are Four Personalities Really Only Two Souls?

Friends and business partners
Michelle Moshay & Peter Teekamp
Carmel, California - Summer 1997

Paul and Mette Gauguin
1885

Paul Gauguin and Mette Gad-Gauguin live in France in 1885. Michelle Moshay and Peter Teekamp explore possible past lives in 1997 in California. In 2007 Peter and Michelle pose in front of Paul and Mette's photograph on a cruise ship to Tahiti celebrating Gauguin's life there. Evidence of similarities between the two couples point to a linkage of souls.

Fig. 5

Chapter Six

Lives That Mimic

Researching reincarnation biographies from a personality-theory perspective has uncovered intriguing findings. The strongest cases appear to involve current lives mimicking past lives in much more detail than other researchers have reported or predicted. This does not mean the present subject has lived a life parallel to that of the historical figure in all respects. But in childhood, and on into adulthood, one finds uncanny comparisons between the two. Common interests, aspirations and talents show up. Deep character traits and behavior patterns surface in similar ways.

A much abridged, two-life comparison of James Madison and James-II illustrates the depth and breadth of the impact a past-life legacy may have on a new incarnation. This chapter includes a sample of parallels that exist between these two lifetimes. A critic may declare any similarity accidental or charge that we are ascribing meaning when there is no causal link. While both explanations may be true for one or more, or even several, of these points, calling all of them chance does not mean they were chance. How many examples are needed before we look for a cause?

The details of this case and the others introduced in later chapters comprise a small representation of their documented evidence for reincarnation and specific past-life connections. Seen as a whole, they suggest that the phenomenon we casually call reincarnation is much more complex and influential than most people think. Start by looking closely at the photos on the cover. The faces clearly mimic one another.

※ ※ ※ ※

James Madison, born in 1751, was the eldest child of a fourth-generation Englishman plantation owner and his socially well-placed wife. He grew up on a then primitive farm later called Montpellier in the Orange County part of the Virginia Colony. In 1939, James-II, an eldest child, was born in another part of the South on a small farm tended by his sharecropper father. It was in such a backward area that its farming methods and economy were not unlike those of eighteenth-century Montpellier.

James's great grandfather John's family settled in King William County, near Hanover County Virginia. James-II's seventeenth-century forebears, six generations before his birth, lived in the latter county. Young Dolley Payne, James's wife-to-be, also grew up in Hanover County near a village still known today as Montpelier.

(In the twentieth century, James-II's wife-to-be taught eighteenth-century English literature at a small liberal arts college in the same Hanover County. She spent six years not ten miles from where Dolley resided for fifteen years. From their hearts in Hanover, Dolley (1794) and Kelly (1982) respectively wrote letters to James and James-II leading to their marriages.)

Though separated by almost two centuries and opposite levels of society, the childhoods of James and James-II had many things in common. In the 1930s Depression-era economy, the sharecropper life of the James-II family mirrored the technical level of a mid-eighteenth century plantation in the western piedmont of Virginia.

With no electricity for lights, running water, radios, refrigeration or heat, and no gasoline driven vehicles, the parents of James-II depended on mules and strong backs for the cultivation of their cash crops. The same hunting for wild game, raising animals and crops, and being self-sufficient for most of life's necessities were essential to both families. Manual labor in the house and fields, annual crops, similar tools and livestock underpinned both the plantation and the farm.

Of smaller stature than their peers and beset with fragile childhood health, the two James created their own entertainment with the help of innumerable cousins and other relations. With access to few books, they treasured those available to them. Each found most intriguing those on ancient history. In devouring them, it was as if both were rediscovering a long-forgotten past.

Education. Both credited key teachers for a solid foundation that enabled scholarly success in college. Each desired progressive institutions of higher education beyond their own region. James's family means made possible his travel north to the early version of Princeton. Given the family finances of James-II, he had to work his way through the state college system. Only in his thirties did he manage to fulfil his own inexplicable childhood expectation of receiving an Ivy League university degree.

Each personality seemed in a hurry to complete requirements for a degree, to get it out of the way in order to focus on his own interests. James crammed during the summer of 1769 to take the freshman exams at the College of New Jersey (later Princeton) in order to save a year. He then compressed the work of three years into two so he could graduate by September 1771. He remained in Princeton through the spring of 1772 to pursue additional studies of his own choice.

After high school, James-II attended a nearby college for two years to insure his rural education was adequate for the big state university. Moving on to the state capitol, he completed his bachelors degree in three semesters. Like James, he then stayed for further study mixing up the courses he wanted to pursue. Both started a lifetime of writing during this period which would deal with the intellectual, social and political issues of their time. Their products exhibit corresponding patterns of thought.

Rhyming Reason. Our mental software affects our patterns of thought, approach to analysis, problem solving and other ways of thinking. The transmission of this factor through soul reincarnation is treated more fully in Chapter Fourteen. However, the soul-mind structure also appears to affect artistic endeavors such as lyrics or poetry. It seems James's soul "never had a poetic bent, for serious work it had been sent."

The previous sentence as an effort to rhyme is on a par with some of the James's and James-II's ventures into poetry. In early Princeton's undergraduate paper-wars, students used all literary forms to make fun of one another. James wrote the following:

Great Allen founder of the crew
If right I guess must keep a stew
The lecherous rascal there will find
A place just suited to his mind
May whore and pimp and drink and swear

*Nor more the garb of Christians wear
And free Nassau from such a pest
A dunce a fool an ass at best.*

James-II, in his twenties, decided to preserve a few serious thoughts he attempted to express in poetry. Self-published, his collection of poems were no better than James's.

*But for restless men
With shoulders that seem to bend
With loads unseen,
To ignore them is to demean
Some dark connate stream
That flows on, with twists and turns,
Murmurs and whispers and calls.
It bubbles near the surface before it falls
Deep in the caverns of the brain,
Perhaps never to rise again.*

After reading the James-II collection, a diplomatic (double entendre intended) friend commented, "... a Shakespeare you're not." Fortunately, both James and James-II decided to stick to nonfiction as their primary modes of written expression.

Political Initiation. En route to Princeton at age eighteen, James visited friends in the then "capital" of the colonies. Philadelphia was the site of growing unrest with British rule. He heard tales of protest actions taken in several of the colonies. He saw patriotic demonstrations. This passage provided his first exposure to the political winds that would stimulate his life's revolutionary course for nearly seven decades.

As a nineteen-year old Congressional intern, James-II met his first President (Dwight D. Eisenhower) and Senator John F. Kennedy for whom he would cast a vote to succeed Eisenhower in 1960. Kennedy's inaugural challenge of "ask not what your country can do for you—ask what you can do for your country" animated him intellectually and professionally for life.

It retrospect, such capital exposures during that formative age to the talk of a nation facing a "new frontier" served as a rite-of-passage for both youth men. Each appeared branded for a life's work in various forms of service to a nation and its destiny. From a reincarnation perspective, such a predisposition may be an aspect of the soul's legacy.

Later chapters show surprisingly similar correspondence of stages in their respective political development through adulthood. For instance, around age forty James was sponsoring the Bill of Rights, opposing a National Bank and urging adherence to the Constitution's system of checks and balances. At about the same age, James-II, unaware of his parallels to James's career, was writing articles and appearing before Congress to push for reform of the federal bureaucracy. He argued for a twentieth-century version of the Revolutionary "Committees of Correspondence" to re-exert citizen control over government and revitalize its system of checks and balances.

Military Service. With no deep sense of career aspirations, James and James-II each received a commission as a military officer in their early twenties, in a time of threat to their country. Both ended their active military service by about age twenty-six.

James received a commission as a Lieutenant Colonel in 1775 in the Orange County militia from his father who was in charge of the county unit. He procured equipment and took part in drills and marksmanship training. He admitted to being "far from the best." James did not serve in the field to become a veteran of the Revolutionary War. His military service was cut short by his election by local freeholders to represent them at the Virginia Convention of 1776.

In the early 1960s, James-II was commissioned a Second Lieutenant in his local National Guard unit (motivated in part by a surrogate-father mentor). Going on active duty he was assigned to administrative positions and reached the grade of Captain. He was relieved of active duty requirements when the President appointed him to a temporary assignment with the potential for a significant career position.

The Sense of Self. After college, neither James nor James-II chose to follow a traditional profession. As self-guided practical scholars, they absorbed knowledge for its applicability to the roles thrust upon them by changing life circumstances. Each saw himself called upon to be of service and was dedicated to establishing institutions of good government. Neither relished political campaigns or political infighting, but answered the call to duty in ways consistent with his time and place, with James becoming one of the Founding Fathers.

Regardless of the role being played in official positions or volunteer undertakings, each would conscientiously assume organizational duties, including note-taking and drafting documents to meet the needs of the moment. Both were "sober in temperament and methodical in manner," committed to making things work as smoothly as possible. Each was reticent to speak in public, but could be clear and forceful when called upon. Both were good negotiators and workers of compromise.

Modest in dress, each avoided standing out in a crowd. Neither tried to be a charismatic figure, although both enjoyed deep conversations on serious subjects. Jealous of their personal privacy, each tried to control the public record of their personal affairs. Each had a "marked penchant for doctoring the records of his life." James destroyed much of his private correspondence and personal papers in retirement. James-II found himself at age sixty-six, the age James retired to Montpellier, purging his own personal archives of embarrassing items.

Lifestyles and Finances. Financial challenges were significant to both their lives. Although one of the most well-off families in western Virginia, the Madisons were subject to the vicissitudes of weather, market prices for their tobacco crops, and the double-edged sword of a slave-based economic unit. While James felt it immoral to survive on the labor of slaves, he could not afford a life dedicated to public service without them. In later years James discovered this system to be unsustainable. His old age was fraught with family debts, misguided investments, and economic turbulence. (His widow Dolley ended her life in penury.)

The years James-II spent in the military and government made it possible for him to insure that his children received a good education and social skills to compete in professional careers and in the society at large. However, when he resigned his career appointment he chose to forgo any future financial benefits. He, as James did after his Presidency, faced difficult choices regarding financing the rest of his life.

Both believed that public service should not be a route to financial gain, and each eschewed the opportunity to use official positions or service for private enrichment. The new American government had no provision for James to build up retirement funds and his salaries also had to cover many official living expenses.

Both men tried to build some equity through investments in real estate. James finally sold most of his during a period of decreasing values and James-II sold his before the real estate bubble of the early twenty-first century. Neither had much interest in financial speculations or the long-term patience to deal with property-investment issues.

Fortunately, James-II was not burdened with a money-losing farm and decided for philosophical reasons to live the second-half of his life in voluntary simplicity. James, having to rely on income from the family plantation, was compelled to take charge of it and its slaves after his father's death. This haunted the rest of his life.

Slaves and Guilt. James had a deeply felt empathy for the slave's life. He sold his personal slave Billey in Philadelphia (for less than his market value in Virginia) at the end of the Continental Congress in 1783. This allowed Billey, under local law, to be free after seven years. James wrote he could not "think of punishing (Billey)... for coveting that liberty for which we have paid the price of so much blood."

Throughout his life James championed the end of slavery, but his actions were ineffectual and his own expressed desire to free his plantation's slaves never materialized. Although promising to free them upon his death, he never figured out a way to make it financially feasible. (The laws of Virginia also made it expensive. While Washington, a good businessman, could free his slaves, Jefferson, like Madison, a very bad businessman, could not either.)

Focused on great social changes, James favored a new colony in Africa for slaves freed in the United States. He, Jefferson, and James Monroe, supported the American Colonization Society (ACS) with the goal of providing a homeland for as many freed slaves as possible. The colony of Monrovia (named after James's successor President

James Monroe) was founded in 1820. It was declared the independent Republic of Liberia in 1847, with Monrovia as its capital.

James-II's father, like other white farmers large and small, used the cheap labor of the descendants of James's and others' slaves for his own survival as he eked out a living on the depleted cotton lands of the South. Seventy-five years after the Civil War, these Negroes still depended on white farmers for an impecunious survival. Through high school James-II worked beside his poor black counterparts picking cotton and making hay stacks in the fields. His extended family's mistreatment of the Negro workers made him an equal-rights advocate for life.

Souls Review? Madison and Jefferson experienced interracial contacts that developed into close emotional bonds with slaves at odds with the attitudes of most whites in the colonies. The reality of slavery and their feelings about it would bedevil each for a lifetime. Always analytical on the issue, neither found a satisfactory personal resolution for their guilt. Under the ACS repatriation program only about 15,000 freed slaves eventually returned to Africa.

Some believe the soul may reincarnate in a manner that lets it directly experience long-term effects of some of its decisions in the most recent life. Have the reincarnations of James Madison and Thomas Jefferson been exposed to the results of their American Colonization Society efforts? It appears that James-II may have attempted to assess the consequences of James's actions.

James-II decided, maybe for reasons not well understood at the time, to have the President send him to West Africa. He worked for awhile in Freetown, Sierra Leone (a symbol of black freedom and self-government) and visited Liberia (a symbol of liberation). He came to believe that, though continuing to suffer in many ways, descendants of slaves in America would not change places with those whose parents were repatriated to Africa by the ACS.

This experience, prior to any awareness of possible soul connections, strengthened James-II's commitment to a personal policy of social inclusion back in the United States. He now feels his time in West Africa may have a soul meaning. If we identified TJ-? and James Monroe-2 today, would they report similar interest in the current effects of their historical decisions?

Love and Intimate Relationships. At age thirty-two, James suffered the loss of his betrothed to another and the first marriage of James-II ended in divorce. James had few reported intimate relationships with women. He was allegedly spurned by Mary Freneau while at Princeton. His short engagement to Kitty Fowler was broken off by her in 1783. No known public evidence hints of further courtships or amorous connections with women until Dolley Payne in the spring of 1794.

In the more gender-balanced sexual culture of a new century, James-II escaped the restraints James had experienced. He married first at age twenty-one, divorced and had a second marriage and divorce, before eventually marrying the soul that he had possibly known as Dolley. In the style of the late twentieth century, James-II strove to relate to his female partners as equally free and self-sufficient partners. He now wonders whether that reflected a desire to avoid the financial and other responsibilities associated with James's eighteenth-century patriarchy.

Travel and Cultures. James's travel adventures were as limited as his sexual ones. Due to his fragile health and preference for domesticity, James did not travel widely. He turned down every offer to travel abroad to represent the American government. The soul's focus in that lifetime was location specific. James's political activities centered in Virginia, Philadelphia, New York City, and Washington, D.C.

College took him to Princeton, New Jersey. He also travelled through parts of Pennsylvania and Maryland. He journeyed in 1784 to upstate New York, via New England states, with the Marquis de Lafayette for negotiations on a treaty with the Indians (and possibly to look at potential land investments). Writing of this trip, he expressed a "curiosity" for seeing more of the world.

Has the soul used its incarnation in James-II to satisfy that "curiosity"? James-II has travelled to more than 100 countries in his life as a government official and private citizen. James depended on books, reports from envoys, and foreign travellers for insight into other cultures; James-II went to see for himself.

Given France's role as ally, personalized by his good friend Lafayette, during the Revolutionary War and later in countering British policies, it was a top priority for James as Secretary of State (including the negotiation of the Louisiana Purchase from Napoleon) and as President. He became quite early a Francophile through personal contacts and Jefferson's encouragement. Because of his pro-France attitudes, James was awarded honorary French citizenship (along with George Washington and Alexander Hamilton) by the Assemble Nationale in Paris on 10 October 1792.

The first overseas trip James-II made for the Johnson Administration was to Paris. He directly experienced the culture James enjoyed from a distance, and enjoyed being feted at receptions and dinners by a few remaining descendants of eighteenth-century noble French families. Offered a diplomatic visa by the French Foreign Ministry, James-II exhibited a ready taste for the cuisine and *joi de vivre* of French culture.

When James-II arrived in France in the 1960s, a new Napoleon type in the "grand form" of Charles de Gaulle had recently kicked the American military's NATO contingent out of France. He had to grapple with Franco-American politics similar to the "on-again/off-again" relationship experienced by James in the nineteenth century. However, speaking and writing French, he enjoyed things French as much as James had in his day.

James-II's career with the U.S. government also took him to other places that had been of interest to James. Overseas travel and experience in a variety of official positions gave James-II a political and management perspective that put him at odds with various Secretaries of State. During the research on his reincarnation case in this experiment, James-II speculated whether the policy conflicts that eventually led to his resignation from public office might have had roots in the soul's multiple-life experiences in government.

* * * *

Reading this chapter, a colleague asked me, "How many other people could you find in the world whose lives have as many or more similarities with James's life as does the James-II life?" I answered that while I did not know, I was sure that some similarities could be found in the lives of many other people. However, I feel it highly unlikely that we could find such a quantity and symmetry in other lives.

She then asked, "What do you say to the potential charge that you only selected the pieces of both lives that do match and left out all the differences?" I responded that I had made no attempt to exclude contravening evidence. In fact, I found very little.

This brief selection of corresponding physical characteristics, mental attitudes, emotional inclinations, and political and personal interests illustrates how one can begin the evaluation of a proposed past-life connection. It introduces some of the types of evidence that suggests the same soul may have influenced two separate lifetimes. After reading further parallels between James and James-II and the other cases examined in subsequent chapters, I hope the reader will consider the research approach set forth in this book.

Part II
....
Developing Hypothesis

The Part I "mysteries" have been interpreted from many worldviews. However, this section assumes they can be best explained from a scientific perspective. It offers a new hypothesis consistent with recent discoveries and emerging trends in physics, biology, psychology, and consciousness studies.

In a universe as complex as ours, anything could be "possible." But to posit scientific hypotheses dealing with a largely metaphysical subject such as reincarnation requires us to work with the plausible. From that perspective, a hypothesis should not contradict natural laws that we already have good reason to accept as valid.

The facts of the previously described mysteries fall in the realm of the known. But an effort to explain them takes us into the unknown. Chapter Seven describes the route taken to arrive at a project approach that bridges the known and the unknown. Chapter Eight follows an interdisciplinary, cosmological approach to reasonably extrapolate from the known to deal with the unknown.

Chapter Nine outlines the growing scientific basis for the concept of a psychoplasm or "soul genome." Chapter Ten postulates how the hypothesized psychoplasm becomes entangled in the conception of a new zygote. It, along with Chapter Eleven, identifies the contents needed in the psychoplasm package to manifest the unexplained evidence reviewed in Part I. Chapter Twelve describes the challenges of obtaining valid and reliable evidence.

Chapter Seven

A Seminal Experiment

This effort to develop an integral hypothesis to account for the scientific evidence collected by psychiatrists Ian Stevenson, Jim Tucker, and Adrian Finkelstein, physician Walter Semkiw, and many psychologists can only be considered a seminal experiment. The hypothesis and its evidence cannot be seen as definitive proof of life after death or the rebirth of souls. It has not yet been supported by independent scientific replications and remains in a pilot-study mode.

The project's goal is to use a psycho-physical perspective to assess the evidence that suggests people today have bodies and personalities based in previous lifetimes. It is made possible by individuals who agree to have their lives closely examined in the process.

Even though the experiment is still ongoing, I believe that after completing the initial stage, its concept and process offer useful insights for reincarnation research. The best way to find out if it does is to put these ideas in the public domain, seeking reactions from many different perspectives. While other people have taken part in the experiment, I must take responsibility for this partial and still evolving report. I welcome critiques from anyone willing to make them.

My preparation for this effort began when I started graduate work in psychology in 1961. At that point I was puzzled that my personality was so unlike all my peers in the small town where I grew up. I knew I was very different from what my community seemed to expect of me. Entering graduate school, I wanted to become an expert in the area of personality development. It now appears in retrospect that I wanted to become a specialist in that area in order to better understand myself. My motive was not that I considered myself a prodigy in any field; I only thought of myself as unique—more like an ugly duckling.

The favored theory in my psychology department asserted our personalities came through learning from social models and the rewards and punishments of our childhood. The problem for me was that in my childhood environment we had had no models or reinforcement for my type of personality. The sense I had of myself was not compatible with what the socialization pressures around me could have created from an infant who started out like a potter's lump of clay.

This unwittingly preparation continued with my graduate psychology courses at Harvard in 1973–74. For my work on a second Masters degree, I became involved with the renowned social psychologist David McClelland (1917–98). His human motivation theory was quite relevant to my long-standing interest in personality development. It posited that three primary psychological needs motivate humans: The desire to (a) accomplish a personal level of achievement, (b) exercise power over others, and (c) have them affiliate with us.[1]

According to David's view, each person has all three needs, but one is usually stronger than the others. Thus, we end up with a profile of varying levels of focus and energy among the three areas. Interested at the time in the theory's application to government workers, I studied various techniques to develop profiles of different groups of people.[2] Trained in his methodology, I tried to develop new ways to understand how a personality develops.

For my research project with David, I hypothesized a fourth social motive. I posited that humans had a "need for interpersonal competence." By that I meant all humans are motivated to learn how to get along effectively with others. I had concluded that we try to master the art of personal interactions (in a variety of modes) in order to fulfill the other three basic needs that animate us. My research demonstrated we could measure a motive defined in that manner.

I did not anticipate it at the time, but, some thirty years later, the technique David had developed to measure people's motive profiles on the basis of their written or spoken words served as the basis for the methodology central to the psychoplasm experiment. As you will see in Chapter Eleven, the personality rating scales used in this experiment depend on inferences about personality traits from a subject's written or spoken words and observable behaviors.

From my work at Harvard I reached four conclusions: (1) Each personality is unique. (2) Its uniqueness can be measured in a way that distinguishes it from all others. (3) A personality's focus and stability cannot be fully explained by environmental or cultural influences. (4) Most scientists assume one's gene pool explains what is left.

However, my small-town childhood and fifteen years seeking a theory that fit the facts had convinced me that it was not that simple. First, it seemed to me that the parental gene pools in my rural village had produced a "cast of characters" that no reasonable person could predict from knowing their parents. Second, that largely homogenous enclave did not provide a mix of social and cultural stimuli to produce the range of personalities my classmates and I created.

My search for a more plausible theory led me to study Sigmund Freud's concept of the unconscious, Carl Jung's idea of a collective consciousness, Edgar Cayce's reading of the Akashic records, and Teilhard de Chardin's notion of a noosphere. Their concepts suggested to me that somehow human minds merge in space to form archetypes or morphic fields that influence future generations. Returning to government service from Harvard, I continued exploring such ideas.

At the time, I thought of reincarnation—if it were real— as simply a soul who invaded the body of a fetus about to be born. The event was only a mechanism for the soul to get a lease on a new life. There were no significant implications for the current personality; it routinely went about its own business as the soul got a free ride.

Years later, when my publisher expressed interest in a book on reincarnation in the context of personal transformation and karma, I was still only vaguely interested in the personal aspects of the subject. I planned my usual research program for a scholarly book on the history of traditional beliefs in reincarnation. It was to be an academic, cross-cultural approach that I thought might appeal to metaphysical and spiritual groups as well as secular individuals.

Starting the research phase, I decided to review the notes from my discussions with Ian Stevenson more than a decade earlier. I needed a scientific update on what had been a largely unscientific field. I reread some of his cases on childhood memories and the physical markings that linked people today with their reputed past-lives.[4] As dramatic as that evidence was, the part of his work that riveted my attention were his comments on how present personalities sometimes seemed to reflect various traits and capacities of the alleged previous personality.

In Stevenson's work, I glimpsed the possibility that reincarnation might be useful to my personality-theory search. I decided to review tape recordings relevant to reincarnation that had been made when I first dabbled with the idea of reincarnation. One was a trance-channel response to some questions I had a friend pose to psychologist Ron Scolastico in 1979. The second covered three 1983 past-life-regression sessions I had done with psychologist June Steiner.

While the tapes had revealed nothing actionable to me, they implied that living many times in roles that contribute to our current capacities makes us unique and powerful from birth. If that were true, the past would give everyone an experience-based foundation for future development.

I speculated whether my precocious behavior could have been due to traits and capacities carried forward from earlier lifetimes. Did reincarnation offer a reasonable explanation for the individuals I had known with extraordinary "blast-offs" from what had been a limited economic and cultural background? Could it explain the examples of people's seemingly odd career choices, and later rejections of them? I began to think reincarnation more complex than I had imagined.

From my studies in psychology, it had become obvious to me that humans are single and integral beings. Everything fits together and what happens in one part, whether physical or psychological, influences all the rest. I could not rid myself of the notion that if reincarnation were a real phenomenon it had to involve all aspects of a human being. I began thinking in terms of a personality-based framework for my book's approach to reincarnation.

The published reports of the University of Virginia Medical School's collection of cases focused for the most part on young children. The subjects had remembered many details, such as names, houses, specific events, how they died, and other facts that had been subsequently verified. Stevenson had taken many children to the locations of the alleged previous lifetime. They were able to identify neighborhoods, family members and friends, homes, and personal articles.

Information about facial features, habits, personalities, and writing style for some cases had been reported. But the published reports did not include enough of the personality factors and longitudinal data I thought would be necessary to test my emerging concepts. I was at a cross-roads in my reincarnation book project.

About that time, our mutual friend Norm Shealy introduced me to physician Walter Semkiw and his book *Return of the Revolutionaries*.[5] Walter had hypothesized that many present-day individuals had incarnations during the founding of the United States. Using a channelled source and his own initiative Walter had identified a large number of postulated past-life connections for his subjects. Among data that he had collected, that which fascinated me the most were the personality areas that had also intrigued me in Stevenson's reports.

From my personality-theorist perspective, the common traits reported by Stevenson, Tucker, and Semkiw had significant implications for both a theory of reincarnation and for human development. They and others had reported interesting anecdotes that stood out in individual cases, but I wondered whether they were fragments of a larger picture. Did reincarnation involve the carry-forward of more than random features, a few memories, and some coincidences?

Given that the present tools of empirical science cannot directly test what goes on outside our 4-D existence, we cannot set up a laboratory in a "6th-dimension" to prove reincarnation. However, since we have data that appears to connect people living today with previous lives, I felt we could test the value of hypothetical mechanisms to account for parallels between the past and the present. I believed some promising methodologies might be developed.

The possibility excited me. If the evidence I saw among some of the cases withstood closer scrutiny, and more strong cases could be identified, it could mean some sort of soul legacy influences all aspects of an individual life. *Whether we recognized it or not, all facets of our lives might carry traces of previous lives, particularly in our early years. The trajectory we start life with might well be predisposed by the past.*

This would mean the course of the current life might be influenced much more by the collective legacy from previous incarnations than anyone had ever imagined. That would make it necessary for us to understand our earlier lives before we could really understand this life. This informa-

tion might be important to anyone concerned with a person's life: parents, teachers, employers, friends, lovers, partners, and voters.

After being introduced to his work, I attended a July 2005 meeting of people Walter had met in the course of his research. It included individuals who believe the souls now incarnated in them might have had previous lives during the American Revolution. Other participants had independent and self-reported cases. Several believed they were members of soul groups with past-life connections that had survived into this incarnation.

Among the independent cases, a number of them included evidence that I felt could be verified by third parties. I was heartened to find that the strongest cases reported at the meeting were much more complex than the published reports. A few were live examples of the concept I had been working on. Some later took part in what became this personality-based study.[6]

Out of that gathering came a "learn-as-you-grow" approach. In this unexpected detour from my traditional, historical-book research project, I had glimpsed a developmental model that might explain the gaps in conventional personality theories that had perplexed me for decades.

Despite its informal and sometimes contentious beginnings, the experiment began to take shape.

Expanding the search beyond the initial group, I found a number of very credible published cases that included personality-related evidence. As I began to read about them and talk to the individual subjects, I discovered even more examples of potential evidence that pointed toward an integral and organic reincarnation process.

These self-identifed cases and others I investigated were consistent with and supported by much of the Stevenson data and some of Semkiw's work. The most robust cases among them clearly pointed to possible connections between two lifetimes in all the areas I had hypothesized. They suggested something like a holographic transfer of psycho-physical patterns from one lifetime to another might be involved.

In each case, all the categories of evidence served to form a whole person. In both the past and present lives, the relevant physical features and personality traits appeared to be organically integrated to produce the sense of one complete being. It appeared that only an singular and integral process involving two lifetimes could account for it.

Given this array of evidence, I designed an Integral Model to identify and evaluate all components of the apparent reincarnation package. I intended it to serve as a building block for an eventual general reincarnation hypothesis. All the cases in the experiment were evaluated by the procedures discussed in later chapters. This book reports on that process and its findings.

***

You can learn more about the ongoing progress of this research by visiting the Reincarnation Experiment web site at <http://reincarnationexperiment.org>. Updated methodologies and new findings are posted as they become available. You may participate by adding your own case to the project's data base. Yours may become one of the "spotlight" cases based on tangible and verifiable evidence. If you have only clues to a potential past life, you can obtain advice through the web site for further testing of your hypothesized match.

Chapter Eight

Self-Learning Universe

"Curiouser and curiouser!" cried Alice in Lewis Carroll's classic children's book *Alice's Adventurers in Wonderland*. We have no better words to describe the amazing new scientific discoveries about the nature of nature. Physicists now proclaim we live in a quantum universe, arrayed in ten or more dimensions, grounded in an energetic plenum of plasma. The possibility of such a multilayered reality makes the notion of ideas like reincarnation seem less strange.

Much of the lack of interest in reincarnation among scientists and the secular public has been predicated on a narrowly defined sense of the universe. As scientists like theoretical physicist Lawrence M. Krauss at Case Western Reserve University show that the standard model is neither complete nor exact, inquisitive minds are pulled beyond the proverbial box.

Although a defender of the scientific canon, Krauss confirms at *eSkeptic.com* that strange behaviors by subatomic particles are only shadows of the real story. He confesses we have no clear and consistent theory of quantum gravity. He reveals, among other things, that we neither understand the nature of dark matter that allegedly accounts for 95% of everything nor the matter-antimatter asymmetry found in the universe.[1]

Many readers are familiar with Michael Talbot's generational reframing book *The Holographic Universe*.[2] Published in 1991, it synthesized research from many disciplines, particularly the rapidly evolving fields of physics, neuroscience, consciousness studies, and systems theory. His simple, yet profound gift of the notion of the universe as a hologram—each fragment containing energetic patterns that represent the whole—has helped physicists and others understand the interconnectedness of entities and the role of information in shaping them.

A Multidimensional Universe

Almost weekly a new discovery declares obsolete a law, principle, or formula that had been considered a foundation stone of science. For instance, the week I was revising this chapter, three world-class laboratories (in Russia, Germany, and Australia) announced that galactic dust forms spontaneously into helixes and double helixes [the pattern found in DNA]. They reported the inorganic creations had memory and the power to reproduce themselves.

For the *New Journal of Physics,*[3] these scientists wrote the self-organizing particles are held together by electromagnetic forces that could contain a code comparable to the genetic information embedded in organic matter. These codes or "memory marks" are involved in self-duplication. About the same time, the U.S. National Research Council suggested NASA should search for "weird life," defined as organisms that lack DNA or other molecules found on Earth.

A NASA search for new life forms would remind us of how little we understand our own. In the human body, half a million cells die each second. Each day about fifty billion cells in our body are replaced. The result is that each year about 98% of the cells in our body are new incarnations. In this renewing process, cellular life is a cycle of "discarnation" (disappearance of a physical cell) and reincarnation (the appearance of a new cell with the same energetic patterns).

The standard model of physics does not yet encompass the concept of an info-energetic mechanism that conserves the self-renewing organism's patterns through this cycle of cell death and rebirth. However, a plausible construct of a psycho-energetic control field is emerging from current research discussed in the following chapter.

As we'll see, it is consistent with biologist Rupert Sheldrake's notion of morphogenetic fields in nature or the old vitalist idea of intelechy, both of which suggest an inherent structure that unfolds and informs the process of development in all species. They suggest all entities may be embedded in entropic information that could be seen as the medium of consciousness.

Transcendent Consciousness. We now have robust empirical evidence that human consciousness transcends the five physical senses. Cleve Backster's biocommunication research, along with others, has demonstrated that human thoughts—undetected in the electromagnetic spectrum—elicit responses in other species (plant and animal) and isolated human cells.[4]

Physicist William Tiller sees this non-physical connection as a psycho-energetic force that transcends the particle/wave reality of our physical experience, with its limits such as the speed of light. He bases this notion on experiments that demonstrate that while particles (the physical realm) must always travel at speeds less than the velocity of light, they are guided by superimposed pilot waves that must travel at speeds greater than light. These waves are known as "information waves" which, in the context of this book, may be the realm of consciousness.[5]

Dean Radin's books *The Conscious Universe* and *Entangled Minds*[6] and my own *Our Solarian Legacy: Multidimensional Humans in a Self-Learning Universe*[7] (in a general way) provide descriptions of research done over the last three decades on the transcendent and interconnected qualities of consciousness. They, too, support the notion of consciousness as a fundamental element of the universe instead of as epiphenomena of single, separated brains.

Consciousness research by a number of people suggests that information, in the form of memory, cannot be destroyed by the warps of space-time. For instance, Gary Schwartz's book *The Afterlife Experiments*[8] demonstrates that memories accumulated during a life can be retained after death in an active, but not-yet-understood conscious form. Information, which can be verified by independent parties, has been communicated from the after-death forms to psychics.

After decades of paranormal research by pioneers like P.M.H. Atwater and others,[9] physicians in the United States and the United Kingdom have taken so-called, near-death experiences seriously. Even the mainstream periodical *Newsweek* highlighted in its 23 July 2007 issue the work of doctors Sam Parnia (British pulmonologist) and Jeffery Long (American radiation oncologist) on the independence of the mind and its memories from physical brain cells and normal states of consciousness.

Parnia says, "We still have no idea how brain cells generate something as abstract as thought." Long seeks more cases of cardiac arrest patients and others who have memories from periods of unconsciousness. He expects many to recall both what happened to them and the medical personnel's conversation during these periods. Some have already reported they encountered the spirits of people they knew who had died earlier but now inhabit another realm.

Such out-of-body experiences (OBE's) do not require that one almost experience death.

Individuals learn to induce OBE's through the use of sound in the Monroe Institute's Gateway program near Charlottesville, Virginia, and with the use of mirrors in therapeutic sessions in Germany. People in these programs and U.S. Department of Defense and CIA remote viewers have reported conscious encounters and communications with nonhuman beings. These contacts indicate that consciousness is not something limited to a physical Earth species.[10]

Professional trance-channels like Kevin Ryerson,[11] and ordinary humans, when using their subtle senses of clairvoyance and clairaudience receive communications from nonhuman voices. Many of the nonhuman entities claim to be permanent centers of consciousness with lives as former humans or other organic beings, but now exist in dimensions invisible to us. Some of them who say they have already lived human lives reportedly expect to do so again.

Considering such evidence and the emerging holographic implications of new discoveries in physics, many scientists no longer think of the universe as composed of only matter and energy. Some, like physicist Jacob Bekenstein, now regard information (as in energetic patterns not unlike Plato's concept of the ideal form) as a "crucial player in physical systems and processes." He wrote, "Indeed, a current trend, initiated by John A. Wheeler of Princeton University, is to regard the physical world as made of information, with energy and matter as incidentals."[12]

The fields of information conceived by physicists include patterns, data, and universal forms, like the dust helix mentioned above. In my view, they may be seen as self-creating imprints in the energetic record of the universe. If we consider matter and energy as "forces" because of their capacity to shape entities, then the information that determines the behavior of the entities must also be considered a force. I cannot claim credit for this idea.

Japanese physicist Shiuji Inomata, at an conference in Tokyo, Japan in November 1996, introduced consciousness into the standard model. Presaging Bekenstein, he predicted the linear reciprocal relationship between mass and energy postulated by Einstein would soon be seen as a triangular relationship where mass and energy also interrelate with a third force: consciousness.[13]

The law of conservation of energy states that energy (and matter by inference) may neither be created nor destroyed (only transformed). If that law is valid, another force equal to those two, such as information/consciousness, must also be indestructible, subject only to transformation. Thus, a local field of consciousness could not be destroyed, only transformed.

Context for Reincarnation. I believe this brief overview of new scientific insights means it is no longer reasonable to believe that the consciousness expressed by humans is completely dependent on and limited to a short-lived, physical body. If consciousness links its physical hosts with other humans and species, it is not unreasonable to postulate that a human-focused field of consciousness (as in the soul genome) could maintain its integrity in a quantum realm.

Such a theory of reincarnation requires a nonmaterial dimension that allows these fields of local consciousness to exist outside of, but interact with our 4-D reality. Present superstring theories' predictions of ten or more dimensions could provide the home for the "unincarnated" phases of such fields. It is likely that scientific acceptance of a general reincarnation theory will depend on eventual proof of a string theory or another multidimensional cosmology.

The second implication of these new scientific concepts is that reincarnation would function as a natural process in a multidimensional, integral universe. It would be governed by the same principles of quantum uncertainty, non-locality, and entanglement that govern all other natural phenomena.

In the microcosm, subatomic particles and photons exist of energy quanta that alternate between particle and wave states. While we do not know how this dynamic actually plays out in a macrocosm, it may provide an analogy for the incarnated (particle) phases alternating with the unincarnated (wave) phases of reincarnation. The view of some physicists that the experimenter's intent affects whether a photon will appear as a particle or wave could apply to reincarnation.

Reincarnation in Nature

This evolving quantum cosmology clearly provides for the possibility of reincarnation, but does nature have a need for it? Would reincarnation meet some of nature's needs? I believe that if natural laws produce the evidence we interpret as reincarnation, the phenomenon must play a developmental role in *Homo sapiens* evolution.

From this perspective, I designed the experiment to test simple and eventually explainable natural principles. I share the view expressed by Thomas Jefferson in 1823: "[Nature's] laws are made for men of ordinary understanding and should, therefore, be construed by the ordinary rules of common sense. Their meaning is not to be sought for in metaphysical subtleties which may make anything mean everything or nothing at pleasure."[14]

If science has revealed anything about nature, it is that nature is economical. In an integral universe, anything created by one process feeds into another process. Nature is efficient. An entity may function for a time, but it eventually gets recycled. Everything serves no purpose not related to its own existence. All is interconnected through all dimensions beyond 4-D space-time.

For these reasons, science today uses the fourteenth-century philosophic principle known as Occam's razor. William of Occam's Latin "*Non sunt multiplicanda entia praeter necessitatem.*" means we should not assume more entities (or hypotheses) than necessary. We have found that the least complicated assumptions are likely to be the correct ones.

Let's assume the three-faceted view of the universe's basic structure (matter, energy, and consciousness) is correct and that all three can only be changed in form, but not destroyed. In that case, there must be a consciousness analogue to the law of conservation of energy discussed above. Reincarnation may be one manifestation of the "law of conservation of consciousness."

Adapting Einstein's formula $E = mc^2$ that describes the interchangeable, but indestructible nature of matter and energy, one could illustrate the interchangeable, indestructible nature of consciousness with a formula like $Mind = Lifetimes^X$. Reincarnation may be the process through which the universe provides for the preservation of learning when organisms—a single-cell bacterium, fruit fly, or complex human being—transition during a "quantum leap" from one physical state to another. But, you may ask, "What kind of universe needs such a mechanism?"

A Self-learning Universe

In the earlier mentioned *Our Solarian Legacy,* I suggest that we live in a multidimensional universe that is a self-conscious and self-adapting organism. I believe it is self-learning because nature obviously experiments with itself, tries new directions, responds to success, and adapts to failure, but always in the direction of new levels of complexity and self-realization. This self-organizing feature, known as autopoeisis, appears throughout nature, including single-cell bacteria that learn how to resist modern antibiotics.

A self-evolving universe needs to take advantage of learning by every generation of all organisms. Species central to the evolution of this universe would develop ways to learn from experience and to improve on their past performance or become extinct. These evolutionary steps are required for "the survival of the fittest." While that concept has been framed in terms of physical processes, it now appears consciousness must also be involved.

At our level, the procedure of individual soul reincarnation would make it possible for a complex, organic universe to experiment and learn from human lives with all their failures and successes. The universe would gain no advantage from a living subsystem (whether cell, plant, or animal) that lost its individual helpful mutations or effective adaptations during the genome transition from one physical form to another.

A self-learning universe would have little need for conscious beings (like humans) that individuate, develop unique knowledge from experience, die, and then have it all dissipate. The evolutionary imperative would be to develop a way to conserve that knowledge and experience.

It would provide for and use reincarnation to make it possible for all life forms to "accumulate, reflect, and take advantage of experience over multiple generations."

This knowledge-conserving process is, perhaps unwittingly, featured at Chicago's world-renowned Museum of Science and Industry. In one quote at the exhibit discussing the role of DNA in evolution, experimental biologist James M. Baldwin states, "Heredity provides for the modification of its own machinery." Baldwin did not elaborate on how the process of selective cell modification is controlled. It may be by non-physical processes described in the next chapter.

In another part of the exhibit, molecular biologist Max Delbruck is quoted as saying, "Any living cell carries with it the experience of a billion years of experimentation by its ancestors." This suggests that cells can modify themselves to some extent in each generation on the basis of their experimentation in that incarnation. The results are reincarnated in the next generation.

These examples at the cellular level may help us understand how the process we call reincarnation makes it possible for a human being (made of billions of individual cells) to change itself on the basis of a lifetime's experience and preserve the changes during reproduction. If nature has developed in a cell the capacity to pass into the future the results of its experience, then it must also have evolved the same cumulative capacity in the cell's human host.

I speculate that if reincarnation were not a part of the nascent universe, some such mechanism would have been developed to preserve learning as a part of the evolutionary process. Chapter Nine postulates a mechanism: the *psychoplasm*.[15]

Chapter Nine

Psychoplasm as Mechanism

The self-learning nature of our universe seems self-evident to me. The trends in science discussed in the previous chapter also seem to point in that direction. New evidence of consciousness in all life forms supports the notion of an impulse in all organic entities—and even in some inorganic ones—to survive, adapt, and express themselves in increasing levels of complexity. In every aspect of nature—from the dust in space to beings as complex as humans—we see tangible preservation of learning and a variety of means to transfer it to new generations.

We see this process at work in Monarch butterflies that winter in Mexico, but die along the way on their return flight northward in the spring. They are replaced by new generations of butterflies that survive the summer in the United States and Canada. Although the new Monarchs have never seen Mexico, they know the way back to their species winter playgrounds. To simply call this process instinct or heredity does not explain the how of the knowledge-transfer process.

We have an even greater conundrum in the "mysteries" described in Part I, where no direct physical or genetic links exist between the past life and the subject who looks and acts like the deceased person. Is it possible that mechanisms like the informational "heredity" of DNA and the "instinct" of butterflies also effect the transfer of knowledge, skills, and physical features to produce the correspondences between present and past individuals described in Part I?

During more than four decades of research, Ian Stevenson was always skeptical of popular conjecturing about the nature of reincarnation. Given that his findings were based on meticulous measurements and indisputable comparisons, he felt there had to be a natural explanation. That led him to postulate an intermediate energy body that he called *psychophore* (which means *mind-carrying* in Greek). He apparently visualized the memories and DNA patterns revealed by his research being transferred between lifetimes in such a container.[1]

Based on widespread research only sampled in the previous and current chapters, I believe Stevenson's concept has considerable merit. It appears that he, at least intuitively, knew about the biofield concept discussed below. I believe he would agree that the entity we respectively label *psychophore* and *psychoplasm* must be capable of encompassing a nonmaterial template that carries forward energetic fields of physical, cognitive, and behavioral patterns. I have used the Greek *psychoplasm* instead of *psychophore* to suggest both the container and its contents.[2]

Supporting the notion that the essence of a human is more than the physical genome, Swiss microbiologist Alex Mauron, in *Science Magazine*,[3] pointed out that "personal identity does not necessarily exactly overlap with [physical] genomic identity." Russian-American scientist Savely Savva and several colleagues postulate in a recent book[4] that something like a biofield control system is required to carry from one generation to another the four fundamental processes of life: development, maintenance, reproduction, and death.

To insure the stability of living entities, Savva believes, the biofield's energetic and information (cognitive) patterns must encompass these four programs. It appears his team's biofield concept is consistent with my idea of psycho-energetic fields and Tiller's concept of information waves discussed in the previous chapter. While Savva's team does not deal with the reincarnation hypothesis, its members' model of an info-enriched biofield appears similar to the psychoplasm idea introduced here.

An Analogy From Biology

The work of cell biologist Bruce Lipton provides an analogy that helps explain the psychoplasm concept. See Figure 6a. He describes the nature of the surface of a cell as a membrane made up of "self-receptors" that respond to the external environment.[5] He postulates that this membrane "chooses" the cell's reaction to external stimuli. He and other biologists believe that no two sets of these self-receptors are exactly alike. Thus, each cell can be considered a separate, self-directing, conscious organism, similar to the concept of a single psychoplasm.

The Human Cell

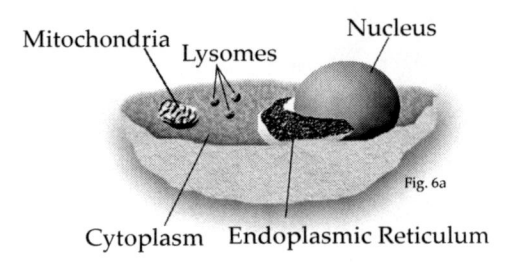

Mitochondria, Lysomes, Nucleus, Cytoplasm, Endoplasmic Reticulum

Fig. 6a

The Soul Psychoplasm

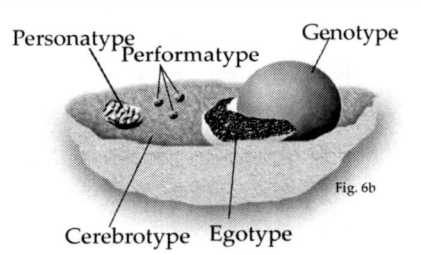

Personatype, Performatype, Genotype, Cerebrotype, Egotype

Fig. 6b

Lipton's image of a self-choosing cell membrane is a biological analogy for the theoretical psychoplasm's energetic field that envelops the soul legacy and protects it through time and space. Like a cell's membrane, an evolving psychoplasm monitors what happens to it and reacts to preserve its unique identity. The cell (comprised of nucleus, mitochondria, Golgi body, vacuoles, etc.) maintains its identity, learns from interactions with the environment, creates memories, and transfers them to the cell that replaces it.

A psychoplasm, like the cell, would have to include the necessary components for its survival, death, and replication. Included among those elements would be energetic versions of the core elements of the human body and its personality. Revealed in the analyses described in detail below, the psychoplasm seems to include an energetic repertory of cognitive patterns, emotions, behavioral styles, memories, and learned skills. See Figure 6b.

In this model, both cells and souls experiment with their environment, perceive what does and does not work, and adapt themselves to changing conditions. With accumulated experience, they learn to successfully coexist with other cells/souls and the immediate organism of which they are parts. With their new learning and skills stored in the individual's package of components, the integral cell and the integral soul reproduce themselves in successive acts of incarnation.

Accounting for Evidence

I believe the psychoplasm/soul-genome concept can account for the mysteries in Part I, the evidence reported by Stevenson and Tucker, and the biometric and personality matches illustrated by the photos and reports in this book. The psychoplasm offers a hypothesis that plausibly explains the imprinting in such cases and adds insight to our present theory of genetics.

In a unique account in the record, Stevenson reports on Lekh Pal Jatav from Negla Devi, India, who was born with mere stubs of fingers on his right hand. He wrote that when an inherited birth defect called brachydactyly occurs, the result is shortened digits on both hands, not just stubs on just one hand. Thus, Jatav's right hand with regular fingers that appeared to have been cut off could not be interpreted as a typically inherited birth defect.

Evidence gathered in Jatav's village indicated that he had spoken of having lived in a place called Tal, had a family there, and had put his hand in a machine that chopped off his fingers.[6] An earlier visitor from that distant village had confirmed a story similar to Jatav's account. Later, Jatav's family visited the village of Nagla Tal where they learned a three-year old named Hukum Singh had in fact lost his fingers in a fodder-cutting machine and died a year later of other causes. Jatav's story about a previous life was further confirmed when he recognized a number of people in Tal and accurately pointed out where the machine had been at the time of the accident.

This case apparently involves a transfer of physical features that, according to our present view of mutations or adaptations involving DNA, would be impossible. Many cases of corresponding physical features—normally considered to be based in parental DNA—have appeared in past and present lives where no direct genetic connections exist between the two families. In the Jatav case, something has to have recorded the energetic imprint of the damaged hand of Hukum Singh, preserved it, and implanted it in the developing fetus of Lekh Pal Jatav.

What Is the Mechanism? The psychoplasm concept expands the current model of genetic inheritance. Scientists presently assume a combination of forty-six pairs of chromosomes (23 sets from each parent) carries forward all the data necessary for the zygote to develop into a complete human being. It is generally accepted that physical traits are transmitted from the parents to their progeny through this genetic package. Body types, hair patterns, predispositions to some diseases, and more are considered gene-based. Even behavioral inclinations are believed by some to be transmitted via the genome.

However, the evidence in this book connecting past to present lives covers more areas than the above-mentioned factors attributed to DNA transfer. Beyond unique physical features, reincarnation evidence includes a variety of personality traits, specific knowledge, and special skills. A reincarnation mechanism must be more comprehensive and capable of activating genetic off/on switches beyond or different from those normally identified with inherited traits.

Is there reason to believe that influences beyond the parental chromosomes actually exist? My hypothesis that a bioenergy field independent of the parental legacy has influence on the developing zygote may be supported by research on twins and fruit-fly (Drosophila) evolution.

Studies of identical twins where one is a homosexual have found that two beings who share enough chromosome DNA overlap to appear as almost 100% physically identical still differ in important respects. For instance, the other is usually not homosexual. This indicates an individual's inherited legacy has space to reflect influences beyond what the parents contribute. Twins researcher N. E. Whitehead writes, "no scientist believes [parental] genes by themselves infallibly make us behave in specified ways.... Genes create a tendency, not a tyranny." [7]

Professor David Houle's Integration of Genotype and Phenotype research group at Florida State University found that natural selection both drives overall patterns and determines to some extent which genetic mutations survive.[8] In my view, this may suggest that something like a fruit fly's biofield has control over which mutations manifest in the next generation. If such a mechanism works with fruit flies, it would follow that a biofield psychoplasm could exert similar control over which genetic patterns manifest in the zygote from one lifetime to the next.

This system of selective and reversible control is known in science as epigenetic modification.[9] In this process, anomalous features show up in an organism even though the basic genome has not changed. This means a defect or positive addition occurs during cell differentiation in one generation and them disappears without apparent effect on the genome. It can also occur during several rounds of new cells even though the basic DNA sequences do not change. Which parts of an organism give these on/off commands is unknown. Perhaps it is in the DNA we now call "junk." Could it be the psychoplasmic patterns that activate the on/off switches?

Stevenson's Jatav case illustrates how an epigenetic psychoplasm could absorb the energy patterns of a wound and then activate the on/off genetic switches to reproduce it in the reincarnated child's hand. In effect, the soul genome implants the modified past-life genetic information as it energizes the merging of an ovum and a sperm into the developing zygote. The result is a single incarnation of the wound patterns developed in the previous lifetime. The child of the adult with the defect will not have that feature. Such is the nature of epigenetics.

Students of the history of evolution theory may notice that the psychoplasm concept resembles somewhat the notion of eighteenth-century French naturalist Jean-Baptiste Lamarck. A proponent of evolution, his concept also provided for the inheritance by one generation of characteristics, attributes, and talents acquired during the previous generation. Overshadowed by Darwin's simpler concept, he was perhaps ahead of his time. The biofield may prove him right.

The Transfer Process. How does a psychoplasm entangle in the gestation of a new body so that the physical features of the previous personality appear in the new incarnation? In his time, Stevenson framed the issue from the duality model usually associated with reincarnation.[10] He mused that the nonhuman personality must select its new parents to match itself, cause its next physical body to choose the right ovum, or directly shape the new body for the desired effects. All three options reflect the idea of a soul selectively acting outside of natural processes.

I consider the psychoplasm to be an as-yet-unrecognized natural entity or process that extends in two directions beyond our current understanding of the genome's role in evolution. The psychoplasm could interact with what is now referred to as "junk" DNA and manipulate the on/off switches that give us the epigenetic effect described above. The psychoplasm could also embed the genome within a non-physical, biofield set of patterns.

Geneticists have solidly demonstrated at a physical level the transfer of DNA/RNA patterns from one generation to the other. These DNA are widely distributed in the body and yet specific enough to be considered proof of maternal (mtDNA) and paternal (Y) links over many generations. Scientists see this process as a purely physical event, but that view is changing.

They must deal with the quesion of how DNA transmits in microscopic bits of sperm and ova the staggering amounts of information necessary for the production of functioning humans. The most likely answer is that the data is energetically condensed and transmitted in wave form. Now that we know DNA are just quanta of energy, one can easily imagine the package of visible DNA being tucked within the higher frequencies of other dimensions.

Recent experiments with the so-called "phantom effect" suggest that genetic strands should be seen as holographic blueprints. With a holographic quality, the DNA patterns can communicate with other energy fields outside their own organism. Russian physicists Vladimir Poponin and Peter Gariaev, with no reincarnation agenda, have theorized that DNA interactions with energy or light fields may occur through microscopic wormholes.[11]

Such a wormhole could be the route through which the psychoplasm entangles itself with the physical reproductive process. Through this entanglement, it would modify the sperm and ovum's holographic DNA patterns to also reflect those of the psychoplasm—what we might call the soul-genome's legacy. Drawn to one another by energetic vibrations not yet understood, the parental genome and this soul legacy interpenetrate and shape the newly forming embryo or fetus.

The Container's Contents

What are the contents of the psychoplasm that comprise this soul legacy? The program at the University of Virginia, developed by psychiatrists with a focus on child development, has collected a range of data that one would find in general medical, social, or psychological assessments. It includes details about physical markings, childhood memories about an earlier life, habits, emotional patterns, and special knowledge and skills. Much of this data has been published and is in the public domain for other analysts to exploit.

On Semkiw's web site www.johnadams.net, he states, "I use three criteria... to establish past-life matches." The criteria are described as facial architecture, personality traits such as demeanor or habits, and membership in a soul cohort (to be discussed later). He later added a fourth: confirmation by a being known as Ahtun Re accessed by trance-channel Kevin Ryerson.

Finkelstein believes his skills with hypnosis make it possible that "past-life regression performed during a somnambulistic state [is] sufficient to prove... reincarnation." [12] For those who do not accept this assertion, his published account of the Monroe/Laird case provides more tangible data. They include personality traits, physical and emotional problems, linguistics and writing style, and phenotype data that are more persuasive than his use of hypnosis.

Self-identified cases in this book like Keene, Kent, James-II, and Kee and other affinity cases are very strong in biographical data and confirmation of memories. With them and a number of similar cases in the public domain, and in my files, I had a significant range of data to undertake a meta-analysis of physical and personality factors reported by the different researchers.

I identified the factors that usually appeared in the most well-developed cases. What I believe is the most comprehensive and coherent picture of factors both relevant and important to reincarnation research emerged from the analysis.

By overlaying a variety of cases, I saw a tangible pattern emerge. It was as if I had stacked transparencies one over the other, each with only a fragment of a photograph, until the whole face appeared. The structure of an apparent soul-based, reincarnated personality surfaced.

The emerging picture included everything from physical features, mental capacities, personality traits, emotional profiles, behavior patterns, specific areas of knowledge and skills, detailed memories, and areas of career interests, among others. If one takes these areas of evidence for reincarnation seriously, one is dealing with a phenomenon that transfers the underlying patterns that predispose an entire personality.

From the array of corresponding biographical data I had from a selected group of cases, I constructed five somewhat mutually exclusive categories that reflected the major elements of the human personality. I have called them the physical genotype, the cognitive cerebrotype, the emotional egotype, the social personatype, and the creative performatype. The five relatively discrete components of the psychoplasm are somewhat analogous to significant elements of a single cell—as illustrated in Figures 6a and 6b.

Combined, they appear capable of conveying the soul-genome's full legacy to a new body. It appears the psychoplasm transmits the five factors as efficiently as the cell does with genetic blueprints of proteins. While some of the traits may be partially latent or express themselves in subtle ways, all appear to make their respective contributions to the new personality.

Five Factors

This book's hypothesis posits that an infant begins life with its past-life legacy. With it as a foundation, the infant interacts with its new environment and social network to create its own unique contribution to the ongoing process of evolution. If reincarnation works as the present evidence suggests, this cumulative, multigenerational legacy, modified by each lifetime of experience, becomes the inheritance of the psychoplasm's next incarnation.

I found it possible to identify and evaluate the five factors in all the cases listed in the Introduction and in other cases initiated through this experiment. Following are brief descriptions of the five factors:

Genotype/Phenotype. The most visible—but not necessarily the most important—factor in a psychoplasm's container is that of physical characteristics. Unless the present and former bodies have out-of-the-ordinary features or a clear deformity, the face is the most obvious area of correspondence.

While a valid past-life identification should show common facial features, similar features do not necessarily mean a definitive linkage. For this reason, the overall phenotypes in both the present and past life must be compared. The physical characteristics considered by biometric science to be least susceptible to genetic mutation or external influences are the body type, facial geometry, ear form, hand and finger shapes, voice, and odor.

Cognitive Cerebrotypes. Psychologists have many techniques to assess an individual's mental style and capacities. Some focus on comparative categories—as in quadrants like sensor/intuitive and thinker/feeler. IQ tests cover a variety of skill areas: pattern recognition, spacial perception, verbal skills, logical reasoning and classification skills. They test one's capacity to perceive and comprehend a set of circumstances and skill to solve complex problems. Others involve knowledge, short-term memory, or specific areas of training.

In reincarnation matching, the goal is neither to discover the parties' specific scores on such tests nor to judge whether they are normal or bright. A past-life identification depends on an estimate of the probability that the mind that incarnated in the past life animates the current one.

Emotional Egotypes. The egotype factor determines how each person copes with the environment. Psychologists have devised untold numbers of measures to distinguish individual differences in emotional traits, types, factors, and behaviors. The reincarnation researcher does not judge which types work better and which are the least effective. He or she simply needs to compare two emotional packages across time.

While the circumstances vary, the basic psychological set should show itself in both periods in history. The scores arrived at should result in the perspective that a hypothetical time traveller might have if he were able to look in on the two people. He should be able to say, "Ah, Mr. X certainly behaves like Mr. Y a century ago."

Social Personatypes. The individual's personatype determines how she or he relates to others. Many readers have undoubtedly been the subjects of various personality tests that purport to give a label that identifies the subgroup to which the person belongs. The five interpersonal traits used here identify patterns that transcend historical and cultural differences.

They provide a basic picture of how the incarnated soul interacts with other souls in different human situations. The personatype scale provides a brief set of descriptors on how the two being assessed handled themselves in public and private circles.

Behavioral Performatypes. All humans regardless of status, health, skills, or circumstance engage in work and play. In the former we may think of vocation or professions and in the latter of hobbies or avocation. Everyone makes choices about the activities that fill their days.

While such choices depend on a society's complexity and technology, even simple human cultures provide a variety of roles that call for different personality types. The performatype scale compares two lifetimes with generic categories that reflect the individual's creative traits.

Alternative Mechanisms

The psychoplasm concept is not the only hypothesis being offered on the basis of the currently-evolving, natural model of the universe described in Chapter Eight. Psychologist-philosopher Jeffrey Mishlove has formulated an alternative hypothesis based on Carl Jung's concept of archetypes to account for the phenomena associated with reincarnation.[13]

Mishlove starts with Jung's notion that primordial archetypes are living ideas that influence thoughts and feelings, and consequentially activate human actions. He posits that if archetypes comprise "organizing structures of consciousness," they could account for alleged past-life memories and perceptions of links "among certain individuals vastly separated" in time, personal history, and culture.

Mishlove's concept of Synchronistic Archetypal Resonance (SAR) implies that learning from the species' evolutionary experience is preserved in archetypal form in the collective unconscious. He posits that individuals may activate an archetype associated with someone who lived in another era when confronted with a surprising set of coincidences or synchronicities. While universal archetypes and a collective unconscious accessible to individual beings are consistent with the psychoplasm model, the SAR concept does not account for the range and specificity of the two-lifetime correspondences discussed in this book.[14]

For instance, the archetype explanation does not account for the specific physical details based in the zygote genotype that appear in both people. The recognition of a synchronicity in adulthood that activates an archetype cannot account for all the childhood correspondences that occur before an individual's recognition of a "meaningful coincidence." In many cases, a third party recognizes evidence of a previous lifetime before an individual is consciously aware of it.

Philosopher Ervin Laszlo supports a concept somewhat analogous to the SAR model in his 2004 book *Science and the Akashic Field: An Integral Theory of Everything*.[15] It posits a field of universal consciousness that includes the preservation of individual memories. Making a well-argued case for the integral nature of a universe founded in consciousness, he hypotheses that an individual's experience of being the reincarnation of a historical person results from tapping into the deceased being's memory bank. He calls these memory banks "holographic vacuum-traces of another person's consciousness" and believes they are accessible from any point in space.

This explanation fails to account for the fact of selectivity. Why would the subject have picked that particular, now-deceased individual's memory pool and not someone else's? Is the connection by chance? If it is not random, did an affinity pre-exist the tapping into that memory? This model does not account for physical correspondences that come with birth, before the adult's act of "tapping," or habits and behavioral patterns that appear ingrained from childhood.

Chance and Falsifiabilty

The fact that one can find similarities between the bodies, personalities, and careers of a dead person and a living one does not prove the latter is the reincarnation of the former. As a psychiatrist colleague reminded me, such correspondences "may have no more significance than those between two living individuals." It would be foolish to conclude that a few common traits means a high likelihood of a specific past-life match.

In fact, the discovery of equally well-matched individuals randomly found in the general population would be capable of falsifying the psychoplasm hypothesis in a manner called for by the scientific method discussed in the Introduction. While such random matches would not prove that any form of reincarnation is impossible, they would refute the Integral Model.

To support the psychoplasm hypothesis, a researcher must reasonably demonstrate that the correspondences in proposed past-life matches cannot be attributed to simple chance or other self-evident causes. An extensive data base, with appropriate statistical analyses, to eliminate all other causes is not yet available. However, the quality of data and the significant numbers of correspondences found in the pilot group used for this experiment merit consideration.

The next two chapters describe the types of data used in the Integral Model and why matching data profiles in two lifetimes calls for a psychoplasm-like mechanism to account for them. I believe the five-factors are comprehensive enough to overcome the charge that many people with total-profile matches could be easily located. As a further check, the methodology requires evidence of specific interlocking activities. That means traits identified in the past and present should also result in identifiable behaviors that reflect them in both lifetimes.

The biographic or personality approach to developing past-life connections is not unlike what an Interpol detective does to establish that the suspect overseas is the same person who committed the crime at home. The questions are the same. What does he look like? How does he behave? What skills does he have? What are his habits? What kind of friends does he have? How does he react to stress? A unique personality can be identified in two countries or two lifetimes.

Chapter Ten

Genotype Entanglements

When I introduce my hypothesis, people ask questions like, "Why are you interested in physical and psychological stuff?" and "Doesn't reincarnation just involve spiritual things?" Or make statements like, "I didn't know the soul affected the body or had anything to do with our lives." My response, "If reincarnation is *real*, it involves your *real* life," takes many by surprise.

However, the physical similarities in each case get their attention. The visual images make the point that something out of the ordinary might be involved in determining their appearance. The first question is usually, "Aren't all physical features caused by our DNA?"

Some cases allegedly involve grandparents reincarnating as their own grandchildren, deceased children returning in the same family, or more distant relatives being reborn within the extended family. In such cases, some of the physical similarities between the subject and the alleged past life could be attributed to parental genes rather than reincarnation. But, that does not explain either the cases where matching features are absent in the parents or where the subject and previous personality are not on the same family-tree.

The previous chapter postulates a reincarnation mechanism for the transfer of physical features. This chapter probes more deeply into how and with which areas the psychoplasm must function to produce the physical two-life doppelganger effect.

What is involved? That DNA from the chromosomes of both parents finds its way into their new-born is universally acknowledged. The first things relatives are likely to say about a new baby include comments like, "He sure has his mother's eyes. Look at her father's nose." Courts agree that a certain amount of overlap between mitochondrial and X chromosome DNA establishes the physical links between parents and their children. However, reincarnation does not depend on whether children resemble their parents.

The psychoplasm described in Chapter Nine posits the transfer of a template that involves all physical characteristics included in the genotype. If we posit the psychoplasm or soul involves itself with only parts of the subject's DNA package, we would also have to identify the third-party "decider" who chooses what to leave out and what to put in.

Such an arbitrary third party is incompatible with the concept of a psychoplasm as a natural function subject to neutral, universal principles of nature. This means the complete set of physical patterns existing at the moment of death would comprise the legacy genotype that reappears the moment of the next conception. This genotype imprints its features among the combined genomes of the biological mother and father as described in the previous chapter.

Facial Features. As noted earlier, the data collected by Ian Stevenson over decades includes similarities in birthmarks and deformities between subjects and their alleged earlier incarnations. Among children, he collected dramatic examples of visible, and clearly unusual, markings and abnormal features that appear to connect one unrelated life to another. He also documented that some ordinary facial and body features seem to carry over to the new life.

In his later years he wrote, "I have become convinced… that in some cases unusual facial features of a subject correspond to similar features in the face of the person (claimed as the previous life)."… "I gave little attention to facial resemblances during the early years of these investigations. What I am able to present on the subject now will have value mainly as a guide for future investigators." [1]

Encouraged by his words, Semkiw moved from Stevenson's focus on unique features to an overall comparison of facial appearances. He paired photographs of his subjects with images of their alleged historical personalities. If they seemed to have a general resemblance and he found behavioral similarities he posited a match. He wrote, "Facial architecture, the shape and proportions of the face, appears to be consistent from lifetime to lifetime."[2]

Finkelstein, building on Semkiw's work, made an additional effort to validate the facial-architecture and other physical comparisons between Laird and Monroe.[3] Adapting Stevenson and Semkiw's approaches, Keene (reporting on his own case) and other researchers have published photographic comparisons dealing with facial appearances.

Attention to the face, the physical feature most often preserved from earlier lifetimes through photos, portraits, and sculptures, is expected. However, a careful look at the faces in popular Elvis and Beatles look-alike contests reveals that general photo comparisons alone are unreliable evidence of reincarnation. The photos at Figure 7 illustrate the point. Even when there are obvious similarities, it does not mean the structure is the same. Scientific research requires a more precise comparison of features to identify possible matches.

"Look-Alikes" and Different Facial Geometry

John Denver   Steve Kern

Facial geometry is based on the genotype's underlying architecture or bone structure. Superficial factors, including hairstyle, glasses, clothes, and accompanying objects may enhance the perception of look-alikes.

Fig. 7

Biometrics and Facial Geometry. Most readers know about the methodology that provides computer-based facial comparisons used by intelligence, military, and civilian organizations for security purposes. You cannot get into a highly secure intelligence facility without its camera positively identifying you as the person whose picture is in the agency's database. To achieve the 99% percent standard required by security, scientists have mapped the face's geometry and its underlying rigid features. See Figure 8a.

Biometrics and Facial Geometry

Electronic measurement of an individual's unique facial geometry is used for security identifications in large corporations and government organizations. (right)

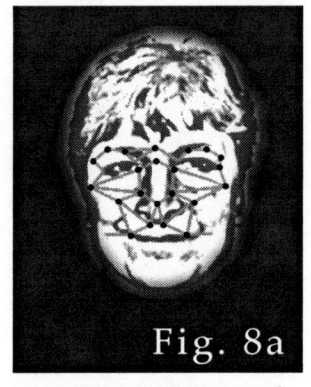

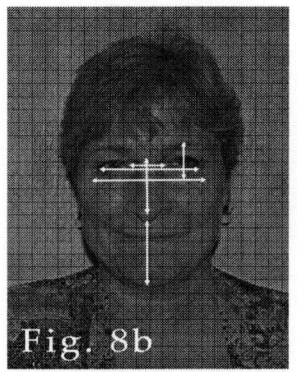

Fig. 8a

Six measurements are made as shown (left), using the grid overlay of a ten-millimeter scale. For comparisons between images with different scales, the six measures are converted to three ratios.

Fig. 8b

The computer uses mathematical models to compare the facial geometry of the individual in front of the camera with a picture in its data base. Software measures the dimensions and relative distances between features such as nose, mouth, and eye sockets. It graphs forehead slope and jaw bone shapes. This data is instantly collated, with comparisons made between the previously secured photograph and the person being observed.[4] If the overall variance between the person and the picture meets the standard, the computer signals a match.

Since most of the previous lives under examination in this experiment do not have photographs, I wondered whether similar biometric techniques would facilitate objective comparisons between subjects and hypothesized previous incarnations where only poor quality photographs, portraits, engravings, or sculptures exist.

In a review of research literature, I found that simple biometric tests had been used to successfully validate that two eighteenth-century portraits painted by different artists separated in location and time were of the same subject. The researchers found the probability that the two portraits were of Wolfgang Amadeus Mozart was high enough to be acceptable in court.

They statistically demonstrated that the two alleged Mozart images could not be the images of two different people. Using data bases of portrait and photograph samples from the general population to compare seven facial features, they calculated that the probability of the two images being of two non-related people was less than one out of ten million.[5]

Based on the Mozart study, I believe that with a few geometric measures we can conduct a similar test of the null hypothesis for past-life matches. That means it should be capable of demonstrating that the similarities between the images of Past-Life X and Present-Subject Y are not the images of two non-related people. While caveats are necessary for comparisons made among amateur photographs, poor quality historical copies, pictures of old portraits, engravings or sculptures, the experimental design I use is a similar process.

With the need for simple procedures flexible enough to deal with a variety of images, I use the six measurements illustrated in Figure 8b. The measurements are facilitated by the ten-millimeter scale shown on the face. In order to make comparisons among the various types and dimensions of available images, the six measures are converted to three ratios. The reader can follow this process on the Facial Geometry Comparisons data sheet at Appendix 1.

The variances between the sets of three ratios measure the degree of asymmetry between the two faces. The smaller the variances, the greater the similarity between the two images. Using the experiment's cases, repeated measurements reveal that the most robust have significantly smaller variances. Cases with weaker personality-factor correspondences have a greater degree of facial-feature variance. Thus, the inter-factor reliability of the model appears to be quite strong.

To follow the process, refer to the photographs of Marilyn Monroe and Sherrie Laird at Figure 9. Measurements of the two photos yield the following ratios: In the first test, Marilyn's eye-ratio #1 is .328, the nose-ratio #2 is .900, and her face-ratio #3 is .667. The same ratios for Sherrie are: #1 = .333, #2 = .900, and #3 = .690. The respective differences between the two women's ratios are .005, .0, and .023. With two variances less than one percent and one at 2.3%, the average difference is less than 1%. A second set of measurements yields respective ratios of .01, .01, and .03. Averaging the two sets raises the estimated variance to a little over one percent.

Two Lifetimes with One Face and One Personality

Marilyn Monroe (1926–1962) Sherrie Laird (1963–)

The biometrics of Marilyn's and Sherrie's faces, as shown in these two photographs, vary by approximately one percent. This suggests that they resemble one another more than either resembles anyone else.

Fig. 9

At this point in the research, I can only say that such a small difference suggests that the two photographs are about as much alike as they would be if they were of the same person. Sherrie would probably be admitted to a secure area by a photo-ID computer with that picture of Marilyn in its data base. Yet Sherrie was born almost a year after Marilyn's death.

Keep in mind, as described in Chapter Twelve, that this is not an exact science. The small number of measurements, use of a millimeter scale, the difficulty of obtaining perfect angles in both faces, and the lack of high dpi in most of the available images sometimes make it very difficult to obtain precise measures. In such circumstances, it is impossible for different analysts to obtain the exact same measurements. While all this results in small, but unavoidable differences in the ratios obtained by different people, the relative relationships appear to remain stable.

As the process is more fully elaborated in its application to other cases in later chapters, its possible use in validation of past-life matches becomes evident. Despite inadequacies in the technique, I believe that with high-resolution images and the right equipment this methodology can be a valuable aid in testing the reincarnation hypothesis and specific past-life matches.

Beyond Facial Geometry. Semkiw wrote, "Body types can be consistent from lifetime to lifetime, though the size of the body can vary."[6] Describing it as only occasional, he wrote, "An individual can have a slight physique in one lifetime and a powerful one in the next. One can be short in one incarnation and tall in another."* He did not include evidence to substantiate these claims. He also noted some correspondences in postures and hand gestures in several cases.

Finkelstein attempted to make other biometric comparisons, including hands and feet, voice, iris patterns, and handwriting. He was unable to secure the cooperation of forensic experts in these areas prior to publication of his 2006 book. Described below, biometric-science research indicating iris patterns are subject to random mutations and that handwriting is mutable by training suggests these two factors are not relevant to tracing reincarnation carry-forwards.

As illustrated by cases throughout this book, the physical correspondences found in the most robust of them go beyond similarities in facial geometry. For that reason facial features are only one of the phenotype factors included in the Integral Model. Additionally, other features are as genetically stable as the basic structure underlying the appearances of the face.

Identification-system specialists select data for their comparisons based on five criteria: uniqueness, universality, measurability, user friendliness, and permanence. The last criterion pertains to the likelihood of random variations during embryonic development. A randotypic feature may be more likely to change over time than a genotypic feature which involves basic structures. I extrapolate from this to assume the most genetically stable features are also the most likely characteristics to survive transfer from one life to another.

Therefore, my selection of features is based on the findings of specialists that exclude the biometric features more susceptible to random mutations or the influence of environmental factors—including training.[7] Facial geometry is solidly genotypic, while fingerprints are randotypic. Handwriting style is somewhat genotypic, but primarily behavioral (training). The eye's iris pattern is randotypic, but the ear form is genotypic. The basic body structure, as in body types, is a manifestation of the genotype.

Based on theses findings, biometric scientists consider the most genetically stable features to be body type, facial geometry, ear form, hand and finger shapes, voice, and odor. The first four of these can be most easily used in research limited to historical personalities. Indirect evidence for assessment of voice similarities exists in some cases. While some small changes could possibly occur in one feature during reincarnation, the overall genotype influence should be clearly evident.

Unique Physical Features. Stevenson's reports and cases reviewed for this experiment reveal that many precise physical factors appeared to be involved in reincarnation cases that do not fit into the above categories. Stevenson—and Keene in his own case—describes wounds to the previous body, some even occurring at the point of death, that correspond to severe damages or deformities in the present body. Other unique birthmarks and deformities that appear to reflect physical wounds or deliberate markings are found in many of the University of Virginia cases.

The physical correspondences between the subjects and alleged past-personalities reviewed in this book confirm the value of comparing such unique features not normally associated with genotypes. Where they can be specifically related to concrete and verifiable events in a prior previous life, these special physical markings are compelling evidence of individual and linear reincarnation.

From Genotype to Phenotype. What is the difference between genotype and phenotype? The genotype consists of the stable set of patterns that are passed from generation to generation. The phenotype is the outward appearance of the genotype that reflects influences during a person's life. In the reincarnation hypothesis, the legacy genome provides a set of basic physical patterns. However, as a child grows, its diet, health, and lifestyle interact with the environment to adapt those basic patterns. Thus, the phenotype may visibly change over time.

Think of the seed placed in the ground as the legacy genotype (the net accumulation of that seed's various incarnations up to that point). The conditions in which the growing plant matures affect its phenotype. This analogy is particularly important when comparing body types. While the basic body type appears to be consistent from lifetime to lifetime, the above factors can have obvious effects on the eventual weight and, sometimes, height of each generation. The rapid increase in Japanese heights after World War II illustrates the effect of changes in diet.

Such changes produced in the subject's appearance mean we may see a bit less than a perfect match between the present and past-life phenotypes. For this reason, it is important to focus on genotype elements that are less susceptible to parental and environmental influences.

Keep in mind that while we measure the superficial phenotype (the result of interaction with the environment), the underlying soul genotype interacts with the parental genetic patterns to establish an important connection between the two lives.

Genotype Entanglement. Although the evidence compiled by a large number of researchers very strongly points to a psychoplasm-effect, is there room in the human genome for such a non-parental influence? Is it possible for external, i.e. extradimensional, input to entangle itself with the parental DNA? If it is possible, it should show up as divergences from the parental genomes.

From recent genetic studies, we now know that the new baby's genome does not consist of simply one-half of its DNA sequences from the mother and one-half from the father. Studies in Canada found that children's DNA may vary from their parents combined genome as much as 12%.[8] These differences—called CNV for copy number variations—consist of multiple copies of some inherited genes and the absence of other genes that should have been transmitted but are missing.

Laboratories decoding human DNA patterns have discovered three other types of variances. The most common are unexplained individual variations called "snips" (SNP for singular nucleotide polymorphisms). The next most frequent are anomalous DNA sequences with obvious insertions or deletions, known as "indels." The other variation consists of "inversions" where the DNA is in the right order on the chromosome, but "upside down."

Craig Venter, the genome-decoding pioneer, found four million such variations in his own genome.[9] This gives a lot of room for non-parental input, some of which may involve secondary messenger hormones that go directly into the nucleus—bypassing the DNA.

Scientists naturally assume these variations to be caused by random breakdowns in the cell reproduction of genes or by physical, chemical, energetic or environmental intrusions. However, Russian scientists have been experimenting with the impact of light waves (as in lasers), various energy frequencies, and even the vibrations of the human voice. They have observed apparent reprogramming of DNA as a result.[10]

All the above described variations have been in the 10% of DNA that is involved in the building of proteins, not in the other 90% which is still called junk DNA. With more than 90% of the sources and purposes of our DNA unaccounted for, the possibility of influence by the subtle intervention of the psychoplasm in the highly sensitive stage of conception is very plausible.

In the not-too-distant future, it should be possible to take the research further and compare some reincarnation subjects' DNA variations with the genetic sequences of their hypothesized previous personalities. Researchers who can identify cases where DNA samples of both parties exist should be collecting them for just such future analyses.

Given the pace of innovation in genome assessment technology, we should soon be able to calculate the probability of psychoplasmic transfer of genetic data, if not find the actual energetic tracks. Technology used to measure psychokinetic impacts on cells through containers that shield electromagnetic influences may help us identify the energetic tracks or soulprints. In vitro fertilization in shielded containers could permit the identification of bio-field impacts by comparison of DNA scans of the sperm and egg with post-insemination DNA scans.

Odor or Scent. A final speculative note on genotype transfers involves a scent and olfactory sensing. Forensic science places individual body odor among the most enduring of genotype factors. The natural odors known as pheromones—chemical particles exuded by a person that stimulate positive responses in the receiver—may play an important role in the validation of reincarnation.

Since scent is one of the most important factors in mutual attraction, the inclusion of pheromone patterns in the psychoplasmic genotype may be important to mutual recognition by members of a soul cohort. If souls desire to reconnect with others whom they have a desire to continue previous relationships, what better way to get attention than through the use of similar pheromone patterns. Keep this playful speculation in mind during the discussion of the soul cohort phenomenon in Chapter Nineteen.

Chapter Eleven

Personality Factors

We humans often think of ourselves in terms of particular traits or characteristics: "I'm an artistic personality. I'm a volatile being. I've always been drawn to animals and nature. I'm a people person and care what others think of me. I live in my head more than my heart. A life of reflection with focus on the spiritual attracts me. I'm afraid to move away from my family even though I don't like them. Give me the open sea and foreign ports. I'm a homebody."

Regardless which of the above, or others, resonate with us, most of us generally assume that our personalities result from the years of experience and knowledge accumulated since birth. But, what if that's not the whole truth?

Self-Awareness. Try to recall an age when you didn't have a sense of your own personality. Even when you used a child's vocabulary to articulate your self-image to someone else, you had a coherent self-portrait. By age two, you had begun to actively define what you perceive to be your personal boundaries and preferences.

At the earliest, babies express the boundaries through "No!" This occurs during the age known as the "terrible twos" when adults attempt to shape them as they prefer. A little later, the self-defined preferences become clear in "My toy." or "My Grammy." statements. A little later, "I want___." and "I can ___." give the child's sense of self a positive definition. Many observers wonder how such complex beings could evolve so quickly from such limited experience.

As people grow older and encounter life-changing circumstances, they may deliberately assume different facades for various situations. However, regardless of how many personas one accrues, when we consciously review them, the sum appears to be more than the fragments. The whole person looking at us from a mirror is more than scattered events and separate memories.

How do we account for the coming together of the multifaceted, multidimensional being we see reflected back to us in the mirror?

Physically-oriented neuroscientists speculate that the brain fills in gaps in our sensory input to create a "sense-of-self," as it seems to do with visual images. Regardless of the location of the synthesizing function, this book tests the hypothesis that a major factor may be the experience of many lifetimes. It postulates that how we come to see ourselves over many lifetimes provides a core sense-of-self with which to begin this life.

Personality Theory

For a psychologist, the concept of personality means a consistent set of psychological patterns that shape the individual's emotions, thoughts, and behaviors. It may also be thought of as the combination of the cognitive, feeling, interpersonal, and creative aspects of being human. Psychologists have developed techniques to define and determine the presence of the traits that appear to be responsible for different types of personalities.

These traits have been tested in psychological studies and proven to be indicative of a person's behavioral patterns. In other words, they reliably predict how a person will likely respond over time in a wide variety of situations. Many of these traits have been found to be operative in the earliest stages of a child's life.

We now have a burgeoning industry of quick personality tests. They range from the use of astrological charts to label different types to the use of adjectives by political media specialists to promote voter bonding with their particular candidates. Others based on psychological studies are more useful here. They include instruments like IQ tests, the Myers-Briggs Type Indicator, AMPM Personality Test, 16-Type Jung Tests, Eysenck Personality Test, Enneagram, and more.

Psychologists design personality traits that are both idiosyncratic and general. They identify basic traits that exist throughout the general population and measure how much of each trait an individual exhibits of each. That makes it possible to create a unique profile for each person which can be used to compare degrees of similarity or difference depending on its use.

The McClelland process discussed in Chapter Seven provides an alternative way to compare profiles of personality traits when it is not possible to directly administer a modern personality test. For his motivation-profile scores, McClelland used samples of written documents produced by or about his subjects. By counting the number of times certain words or phrases appeared, he determined the relative strengths among an individual's motives in order to compare one person to another.

Using the theory behind McClelland's scoring system, I constructed rating scales that would measure the four personality factors introduced in Chapter Ten. The cerebrotype scale identifies the cognitive features that determine how one thinks about questions and how to solve them. The egotype scale deals with the strength and orientation of the basic ego structure. The personatype scale captures modes of interaction with other people. The performatype scale identifies preferences for creative activity that can earn a living or offer rewarding diversions.

To insure the validity of the rating scales, I based them in a long-standing body of research that reduced redundancies among hundreds of personality descriptors into sixteen factors (known as Raymond Cattell's Personality Factors). I integrated the logic of other psychologists who had subsumed those factors into the so-called Big Five Factors. Due to the importance of childhood evidence, I also incorporated the Nine Temperament Traits, developed by Doctors Chess and Thomas and used widely by psychiatrists.[1]

The resulting four scales are used in a two-stage process to compare subjects with their alleged deceased personalities. Using data from self-assessments, psychological interviews, or inferences from biographical data, the researcher can rate each subject on each factor. The second stage requires use of historical or biographical information to rate the past-life personality on the same scales. In this manner one can compare fundamental traits in people separated by lifetimes.

The beauty of the scales is that they do not require an individual to take a test in order to arrive at an accurate picture of his or her profile. Comparable ratings can be inferred from the observed behaviors of an individual or through the use of data from diaries, letters, records of interviews, biographic information, and other documentation.

During the experiment, this approach has proved valuable in

several ways. That definitive personality links can be established between two lifetimes generally strengthens the case for reincarnation. It can also be used as a tool to test the reliability of claims to a past life based on other sources. It may prove to be a therapeutic tool by providing personality insights based on possible historical experiences that can facilitate a more conscious process of self-actualization.

Cognitive Cerebrotypes

We do not have direct evidence that a psychoplasm, independent of but interacting with the physical sperm and egg, actually imprints in the zygote the structure of a previously well-developed mind. However, we do have proof that a full-sized physical brain is not necessary for a human to possess a high-level IQ and other cognitive capacities necessary for a normal, and even exceptional, mental life.

If such a non-brain-based human capacity can be shown to exist, it would mean the cerebrotype concept does not contradict natural law. Fortunately, we have evidence to prove that in the absence of a fully developed physical brain, high-level cognitive functions do continue to operate. Noted parapsychological researcher Stephan A. Schwartz, author of *Opening to the Infinite*,[2] reports on some of this evidence.

Introducing an article by Roger Lewin, Schwartz wrote, "When I was about 12 or 13, my father, an anesthesiologist and professor of medicine, and I were having lunch at the "Doctors' Table" near the hospital where he worked. One of his best friends, a surgeon with whom he frequently operated, got deep into a conversation concerning a young woman who had been brought to the hospital after a terrible car accident. When they opened the skull to relieve pressure they were stunned to discover that in place of her "brain" she had a sac of fluid. Only the brain stem was present. Yet, she was a cheerleader, an honors student, and about to go to Smith.

"Over the years, I would occasionally read about something similar and track it as far as I could get. By the 1970s, when I quit, I had fourteen cases. Like Savantism, these reports suggested to me that the mind being the brain alone simply doesn't work, either theoretically or in practice. This physicalist view lacks that aspect of the self that exists outside time [and] space: nonlocal consciousness."[3]

Lewin reported research by the late neurologist John Lorber at Sheffield University (U.K.). One case involved a bright student with an IQ of 126, but with virtually no brain—one millimeter of cerebral tissue covered the top of his brain stem instead of the normal 4.5 centimeters— about to graduate with a mathematics honors degree. Professor Lorber eventually identified several hundred students with very small cerebral hemispheres who appeared to be perfectly normal individuals, many with high IQs.[4]

These cases demonstrate that only a small bit of "grey matter" is necessary to process prodigious amounts of data. Knowledge is not stored in a chip-like computer memory. Our brain seems to act as the processor through which data is transferred to and from someplace else. Memory is everywhere in the brain, but not really anywhere. One part of the brain can pick up for another when it is damaged or cut away.

Given this and the research discussed in Chapters Eleven and Twelve, it is not unreasonable to assume the mind is more extensive than, and outlives, the physical brain. What I have called the cerebrotype appears to be the essence of this extended mind that appears to transfer from one life to another. As in a genotype carry-over, where physical patterns of the deceased are preserved, the cerebrotype preserves cognitive data. (See Appendix 2.)

Its cognitive patterns incorporated into the cerebrotype shape the aspect of consciousness that reacts to and processes information, whether from external sources or generated within. They affect our analysis, synthesis, and interpretation and how our thoughts are exhibited through speaking and writing. The evidence also suggests the soul's cognitive legacy includes fields of knowledge and innumerable detailed memories of people, places, and events.

Just as IQ scores can be increased slightly by various learning exercises or new experiences, the cerebrotype is expected to progress from lifetime to lifetime. When a previous life reshapes its inherited accumulation of learning from multiple lives, we should be able to see the impacts of that learning in the current life.

The Cerebrotype Factor Rating Scale (included in the case evaluation package provided on the Reincarnation Experiment web site) measures the level of correspondence between two lives even when separated by time. The number of similarities provides a measure of the level of confidence one can have that a case involves the same transcendent mind.

Inferring Cognitive Traits. In the absence of nineteenth century measures of cognitive profiles, the Gauguin/Teekamp case illustrates how the researcher can infer cerebrotype factors from long-term behavioral patterns. The creative work of both obviously demonstrates an intuitive and aesthetic approach in their thinking. The lifelong dedication of both personalities to artistic careers points to independent thinking. Their up-and-down, difficult financial situations reflect undisciplined mental attention to such details. Their frequent moves, relationship changes, and evolving artistic styles show an experimental, self-indulgent mentality.

This process can be further illustrated with the related case of Gad/Moshay. These personalities, in different centuries, respectively support the careers of Gauguin and Teekamp and also share a cognitive profile. Mette stayed behind in France, encouraging Paul to pursue their vision of preserving Tahiti's dying culture in his paintings. She undertook the marketing of his paintings (organizing a Gauguin exhibition in her native land of Denmark) and the support of their children to make possible his gift of art to the world.

Michelle, of French-Lebanese origins, offers her mental and working style to promote Peter's art (and his connection to Gauguin) as a gift to humanity. In her current life, she exhibits the same cognitive profile as Mette. Both are bright and quick, able to combine left (practical) and right (imaginative) brain capacities in promotions, exhibits, and other goal-oriented activities. Both exhibit the attention to detail their more artistic compatriots lacked. They are conscientious and dutiful in support for their artists, while being flexible about their own priorities.

These assessments of similarities between two lives do not involve judgments about an approach. In terms of our reincarnation research, the important point is that Paul and Peter and Mette and Michelle have corresponding scores on all five cognitive factors. The same respective male and female mindsets seem to be acting in two different centuries.

Emotional Egotype

According to the psychoplasm hypothesis, souls have emotions with at least the same range of feelings all of us experience. Their emotional package appears to neither dissolve with the physical body's dissolution nor get left behind when the psychoplasm reincarnates. The best evidence in the most robust cases from several researchers indicates a being's emotional patterns remain largely stable from one lifetime to the next. This suggests we are born with a pre-existing emotional legacy, probably shaped by experiences accumulated over many lifetimes. (See Appendix 3.)

As we look at ourselves or another, we see manifested in daily life the energetic patterns of the emotional state of a soul. If we observe a person with difficulty living up to his or her professed standards, it is likely the soul still has more to learn. Where someone plays with nuances of feeling in emotional expressions, we probably see a soul who is making progress.

Humans continue to live with that inherited legacy, usually suppressing conscious awareness of it, until choices are made to engage in learning experiences that change it. Changes in a soul's emotional legacy from one life to another appear to be made on the basis of insight from experience. If an incarnated soul doesn't learn, it repeats the patterns.

The egotype—as basic as the genotype is in physical matters—consists of all the major aspects of the soul's current level of emotional development. The hypothesis does not predict perfection on any factor. It anticipates a range of emotions in how any individual handles the challenges of daily life. If humans were paragons, we would probably already have, as the Buddhists believe, escaped the wheel of continual rebirth.

Basis for Factor Ratings. Reliable information on how the subject and the historical personality act in a variety of situations makes it possible to determine where they would fall on the four scales. For the most objective results, the ratings of the two personalities should be coded independently to avoid the possible tendency to look for compatible data.

A well-documented Stevenson case illustrates how personality inferences can be made from reports of individual behaviors. Young

Sukla in India was alleged to be the reincarnation of the deceased mother (Mana) of a now older daughter named Minu. Before talk of reincarnation, Sukla, at a little over age three, objected to eating with her non-Brahman family.[5] It was later learned that Mana's family was Brahman. Sukla was always more religious than her biological siblings. People acquainted with Sukla and Mana verified a stubbornness (self-confident) in both.

Sukla shed tears when she first saw the sewing machine owned by Mana and which was very important to her. While still a child, Sukla emotionally related to Mana's daughter Minu in a maternal manner even though Minu was much older.[6] (Sukla rates emotionally warm on the egotype scale, but additional information would be required to rate Mana.)

Evidence that unresolved emotional burdens carry over into the next life comes from the case of Ravi, also among Stevenson's Indian cases.[7] He had apparently been murdered in the previous life by a barber and a washer man. When his case of reincarnation came to light as a child it became known that he was afraid of barbers and washer men. (He would be rated as anxious and worried on the egotype scale.) He later learned the accused murderers, who still lived in a nearby village, were from those two occupations.

In another Stevenson case, Jasbir of India, the newly reincarnated former Brahman, refused to eat anything but a Brahman diet. His new family was of the lower Jat caste. His emotional impulse was so strong that had accommodations not been made, he would likely have starved to death. He refused to speak to people with whom his previous family had quarrelled.[8] (He would be rated as depressed, subject to longer-term evaluation.)

As in the cognitive scale, the larger number of egotype correspondences, the greater the likelihood of a valid past-life connection. However, any given factor rating must also be evaluated in the context of all other factors to arrive at an overall assessment of the strength of the case.

Social Personatype

I suspect that most people who think about the soul conjure up the vision of an ethereal being flitting about in other realms with universal knowledge and cosmic-quality thoughts. Souls are often described as pristine aspects of a godly perfection, without warts or blemishes of any kind. It is life on Earth that tarnishes a shining vessel.

I believe these idealized stereotypes are more projection than reality. Instead of saying someone's "personality has soul," it would be more appropriate to say that "soul has this personality." When one sees the personality, one sees the soul, and vice versa.

In an integral model of reincarnation, human personality styles are the manifestations of an infinite variety of souls. Some are closed and fearful. Other souls are open and optimistic. Some are shy while others are gregarious. Some frown, others smile. Organization types don't like rebellious souls. Happy-go-lucky souls don't enjoy pessimistic ones. (See Appendix 4.)

Persona. Carl Jung saw the persona as the mask or image we present to the world. In the context of the reincarnation hypothesis, it is the outward manifestation of the personatype. The personatype consists of patterns that shape the individual's interactions with others—in sum, a style of interpersonal behaviors. Polar opposites like "introverted versus extroverted" or "cooperative versus antagonistic" are used to label different types. Personatypes may be described by traits like liveliness, social boldness, privateness, vigilance, and self-reliance.

The hypothesis tested here predicts that if a psychoplasm has incarnated in and animated two different lifetimes, we should see imprints of the same personatype in both. That does not mean the two people will have exactly the same social reputation. However, objective observers should be able, with a high degree of reliability, to determine if the subject's basic personality generally matches that of the historical person.

As a phenotype evolves through adaptations to the individual's inherited genotype, the personatype also changes through the process of conscious evolution in a single lifetime. While the essence of the inherited personality should be visible, the researcher will note small differences as the subject reacts to new circumstances and learns to adapt, or does not learn to adapt.

How long it takes to produce readily visible adaptations obviously depends on a variety of influences. However, it is reasonable to assume the individual today does not act exactly like his previous personality did any number of years ago. Since the preceding past life provides the personatype legacy for the current life, the experiment focused on an alleged immediate past-life match where possible.

The Dolley case serves as an illustration of this process. Raised in a socially low-key Quaker household, Dolley had limited exposure to cosmopolitan society. She was taught to wear drab clothing, prepare and eat ordinary fare, and behave in a modest fashion with boys and people outside the Quaker community circle. The ideal was "no extremes of grief or exhilaration."

As she grew older and her family moved to Philadelphia, Dolley rebelled against these controls. She was criticized for her gay colored dresses, the cut of her gown and the shape of her cap. Her participation in Quaker meetings was social, not religious. A decade later she complained to her sister Anna that a Philadelphia stay "brought to mind the time when our society used to control me entirely, and debar me from so many advantages and pleasures."

Kelly was born into a similar traditional society where she learned the norms of her class. However, as the twentieth-century women's liberation movement began she found herself free of much tradition. Having dreamt in childhood of being mistress of a sophisticated parlor (later identified as an eighteenth-century one), she knew she had the power to make her own rules.

An innate sense of style gives each woman a measure of self-confidence in her social touch as she enters new social circles. Regardless of inner doubts (admitted to by both), each exhibits a "natural vivacity" and a magnetic quality of charisma that draw people to them. Their connections with people are not transitory; both maintain lifelong friendships and stay in continual contact with a wide circle of acquaintances.

When the widow Madison returned to live in Washington after James's death, she was always invited to more social events than she could manage. When now retired Kelly returns to visit the town where she formerly worked, she is overwhelmed by a multitude of invitations.

Quick learners in capital cities, when an occasion arose, each naturally accepted the challenge. Dolley easily became the "perfect hostess" for the nation's new President's House in the time of widower Jefferson's Administration. Her sense of fashion and a democratic style established the role of America's First Lady. Later in James's Administration, she arranged weekly "drawing-room levees" for guests from all walks of life in Washington. At Montpellier she managed a household for social events that equalled those of Washington.

Supporting her professional husbands' public careers, Kelly exhibited similar traits of style and warmth in entertaining their many colleagues. She treated everyone to the same gracious acceptance and a genuine interest in their views. All went away with the notion that they had received her special attention. As Dolley did, she delighted in furnishing and decorating her homes and in the serving of cosmopolitan meals and drinks. Her by-words are "elegant simplicity."

Creative Performatype

Vocational counselors help students identify appropriate college majors or professional training on the basis of an interests assessment. Have you wondered about the questionnaires that purport to identify career interests of which you may not have been aware? More often than not, they magically resonate with our own sense of self. In fact many psychological studies have shown that they are fairly good predictors of both career success and personal satisfaction.

The combination of our physical attributes, cognitive capacities, emotional patterns, and personality traits seems to correlate with areas of creative interests, knowledge, and work skills. The reincarnation hypothesis suggests that this clustering of interests is not lost by physical death. As the ways an individual learns to think, feel, and act become part of the psychoplasm's legacy to the new-born child, so do occupational interests accrued in previous incarnations.

Paul Gauguin, the great French postimpressionist painter, once reportedly stated, "When the physical organism breaks up, the soul survives. It then takes on another body." We now believe that Peter Teekamp may be the current body that the soul formerly in Gauguin has chosen. Their similar painting skill and styles, now obvious to us, surprised Teekamp when exposed to Gauguin's work late in his own career.

Personality Factors 105

Soul's Artistic Talent Inherited in Reincarnation!

Gaugin
Age 43

Teekamp
Age 16

Gaugin
Age 41

Teekamp
Age 22

Gaugin
Age 40

Teekamp
Age 26

Gaugin
Age 46

Teekamp
Age 22

Contemporary artist Peter Teekamp, the possible reincarnation of the artistic soul who incarnated in the person known as Paul Gauguin, appears to have inherited the Gauguin psychoplasm's legacy. While Gauguin's artistic talent matured later in life, Teekamp was painting in a similar fashion much earlier. Teekamp was not aware of the similarity of his work to Gauguin until his friend Michelle Moshay discovered his sketches shown above and compared them to the work of Gauguin.

Fig. 10

While Gauguin began painting part-time at age twenty-four, to become a full-time painter in his thirties, Teekamp received recognition for his drawings by age five. In his early school years (while still in an orphanage) he won art contests. Classmates praised his work, giving him encouragement to express what was obviously an innate talent at birth. Teekamp's precocious art, appearing much earlier than in Gauguin, seems to demonstrate an inherited artistic skill. See Figure 10, previous page.

That they share the artistic performatype is clear from seeing Teekamp's paintings along side those of Gauguin, but their lives also share other vocational correspondences. Enterprising, each spent a part of his life in business. Gauguin served in the Merchant Marines and the French Navy. He then became a stockbroker to financially support his part-time painting. Doing what he thought best to support his career as an artist, Teekamp pursued a youthful business degree.

Each exhibited an early willingness to take on a variety of jobs in order to follow their artist inclinations. In addition to his maritime jobs, Gauguin labored for a while as a digger in the Panama Canal project. Teekamp has worked in the retail sector, run a trucking business, and owned a gallery to support his art work. These two incarnations share the entrepreneurial and realistic traits as well as the artistic trait.

A Taxonomy of Creative Interests. How do we know that vocational or avocational desires reflect the interests of the soul? Fortunately, we have decades of work by psychologists (including the pioneers John Holland and E. K. Strong) who ferreted out the links between our core values and interests and the occupations we find most satisfying.[9] While the specifics of jobs and professions change over time, their personality requirements are enduring. The traits that lead to different kinds of work or creativity can be found in all societies in every era. (See Appendix 5.)

From among several taxonomies that separate work activities, interests, skills and values into mutually exclusive categories, I chose to work with Holland's codes.[10] He identified six generic types that transcend genders and cultures, and, I would submit, lifetimes. His types apply to an infinite variety of occupations. Because they represent enduring patterns in the psychoplasm, they are more valid for life-to-life matches than the specific job one might hold.

While not exhaustively comprehensive, the six categories encompass most of the ways people earn their living or otherwise identify their creativity and worth in the world. In this context, Holland's codes take on the fuller significance they merit. They are not job titles or even career categories. As performatypes they suggest the psychological power of deeply rooted values and interests that are nurtured and honed over many lifetimes.

They range from an "artistic" personality with its creative independence to the "conventional" type who follows the status quo in an orderly and reliable manner. The list also includes the "enterprising" type, people who take risks, for their own benefit, to create and promote new products, services, and ideas. It has room for the "realistic" type who works with his hands and body on practical challenges, honoring tradition and common sense.

The last two categories include the "investigative" and "social" types. The first group is very curious, always investigating and analyzing new areas of knowledge. The socially-oriented group nurtures individuals and builds communities.

When one compares two personalities with this set of creative traits it is possible to perceive similarities that would normally be masked by different educational, class, cultural or historical circumstances. For instance, in the Gordon/Keene case, both volunteered to be of service to their respective communities in an institution that suited their common psychological need. Gordon opted for a military career. Keene developed a career in the fire department. With the same need, both sought service in a hierarchical, disciplined, and yet risk-taking profession.

Predictive Capability. A test of the personality-factor model of reincarnation is its ability to predict from the data in one lifetime what kind of behavior will be found in the hypothesized matching lifetime. The prediction should work in both directions, from either lifetime.

The case of Sukla discussed in the above egotype section illustrates how a prediction can be made about how the alleged previous personality would act. Although a small child when Sukla recalled the life

Mana, the model would predict that she would relate in a maternal fashion to Mana's living children as Mana had done. In fact, Sukla did emotionally relate to Mana's daughter in a maternal manner even though Minu was much older. A report of Mana's maternal relationship with Minu before her death could validate the prediction.

Going in the other direction, the model predicts the personality traits to be found in TJ-? on the basis of Thomas Jefferson's personality. Thomas wrote, "In America, no other distinction between man and man had ever been known but that of persons in office... and private individuals. Among these last, the poorest laborer stood on equal ground with the wealthiest millionaire, and generally on a more favored one whenever their rights seem to jar."[11]

Thomas acted upon these principle of egalitarian relationships. His Presidency was epitomized by his *Cannons* (sic) *of Etiquette*, set forth at the beginning of his Administration to govern the protocol of government and diplomatic meetings. His antithesis of the royal customs of Europe was designed to embody a new, American republican model of social intercourse. He eliminated differing treatments associated with rank, titles, heredity, wealth, or social status.

The same ego trait should appear in discernible ways, regardless of circumstances, in a new incarnation. A researcher should be able to identify the same basic patterns in all four personality factors in Jefferson and his reputed reincarnation. According to the Integral Model, a valid case should exhibit the same soulprints in each lifetime.

An Integral Personality. This chapter describes the fundamental elements of the human personality. It illustrates how to use the factor rating scales and biographical materials to compare the personalities of two people separated by time and space. However, its value to the process of reincarnation research depends on the quality of the information used.

For this reason, the next chapter focuses on the sources of information and the methods used to obtain it. Without confidence in the integrity of the process, the subjects and the public will not likely find the results to be credible. And even worse, we may be fooling ourselves.

Chapter Twelve

Sources and Methods

A jury has to assume that the material presented in court is true and represents all the relevant evidence. But, what if it is not? If you were a juror, would you be more confident if you knew the sources of the evidence and the methods used to test it before a suspect was charged with the crime. Why would a prosecutor keep such information secret from the judge and jury?

In a similar context, the CIA, FBI, and others claim they cannot disclose their "sources and methods" to the entire Congress in order to protect their spies and techniques. However, by preventing access to this information, they also prevent Congress from knowing the weaknesses underlying their testimony and recommendations. To make it possible for the reader to take into account the value of the experiment's sources and methods, they are revealed in this chapter.

Alleged evidence of reincarnation linked to two specific lives comes from many sources, in various levels of quality. Sometimes it has passed through several sources. Each one of those original sources and its channels or reporters have their finite areas of knowledge and particular worldviews. We need to know how to take advantage of all of them without somehow falling into the trap of assuming that any one source has provided us access to the ultimate truth.

Human Sources

Random contacts, friends, or family members provide people information about their possible past lives. They share intuitive insights, dreams, or meditations about the subject. For instance, for years before I approached their reincarnation case, Darlene Mettler had been telling George that he acted like Samuel Johnson. Clues received in this manner may also contribute to further, in-depth research involving scientific methods of investigation—as in the Teekamp case.

In the 1970s while exhibiting his work at art fairs, two people he had never met before approached Teekamp. One said, "You lived in France in a previous lifetime." Another said, "You were a Paul in a past-life." After a religious ritual in 1979, his wife, who was not interested in reincarnation, exclaimed, "You are the reincarnation of Paul Gauguin." That pushed Peter to compare photos of Gauguin and photos of himself—the resemblance forced him to reflect.

Based on this, he did a past-life regression session. While he got nothing specific about Gauguin, the details of his "recovered memories" motivated him to learn more about Gauguin. He discovered that Gauguin had, just as he did in his own art, hidden faces in some of his paintings. In 1988, a friend to whom he had told nothing of his Gauguin interest shouted in a frustrated way, "You are the reincarnation of Paul Gauguin." Needless to say, he began to take it seriously.

We should remember that all such hints by others do not turn out as Peter's did. People may offer friends or acquaintances their ideas about possible past-lives in order to please the person. They may want to give the impression of being in the know, perhaps to affirm their reputation as intuitives. Some may make statements to others that reaffirm their beliefs, well-founded or not, about their own prior incarnations.

Self-Reporting. Information about possible previous lives comes unbiddenly, and often undesired, through intuitive flashes of insight, dreams, meditations, or spontaneous memories. How seriously can we take this form of evidence? We know humans reshape their own memories or descriptions of events. Self-reporting poses verification challenges to researchers.

The Monroe/Laird case described by Adrian Finkelstein includes an example of this. During their first contact, Laird reported to him in 1998 that she had a 1992 past-life memory of Monroe's experience of dying. She also related that she had told a physician she thought she was Marilyn Monroe and had recognized Monroe's face as her own in the mirror. While Laird believed them, Finkelstein knew that he had to seek corroboration of her statements.

Finkelstein did not always secure documentation of secondhand information. He wrote that Ted Jordan (by then deceased) had told him that Laird recalled in a telephone conversation details of his intimate relationship with Monroe that only Marilyn could have known.[1] He forgot to have Jordan write down the details to compare them with Marilyn's diaries. Jordan died before they could be documented. Such experiences help researchers improve their methodology.

Hypnosis. Material gathered through hypnotic regressions is a form of self-reporting. It is subject to the distortions inherent in all individual reporting. We are not able to identify and monitor the filters created by assumptions and psychological states in the subconscious that bias communications from that level of awareness. People who experience various hypnotic sessions become aware of different levels of the personal volition expressed in them. In addition to the subject's own unrecognized filters, each hypnotist has a unique approach that influences the nature and flow of communications. Corroboration through other means is required.

Finkelstein believes that he has the skill to direct Sherrie Laird to suppress all her memories of anything related to Monroe except for the "true memories" of her incarnation as Monroe.[2] Anything Laird says during the hypnosis session, he will accept as coming from Monroe's life. However, he recognizes not everyone agrees with the validity of his assumption. Therefore, he supplements his technique with physical and personality data.

Spontaneous Childhood Reports. Some of the most compelling evidence comes from self-reporting by young children. Material of this type collected by Stevenson, Tucker, Bowman, and others is spread throughout this book. Much of it is spontaneous; words burst out in response to someone's unintentional comment or action. I heard a three-year old, when being coerced to eat a hot-dog she didn't want, say, "I did eat hot-dogs when I was boy." Was her previous life a boy?

When a parent is attentive to a child's unexpected reactions to events and gives the child an opportunity to explain, some riveting stories come out. Carol Bowman published her son's first comments indicative of possible past-life memories. They came in response to a friend and therapist asking Chase what he saw when he heard loud fireworks that scared him.

Five-year old Chase responded, "I'm standing behind a rock. I'm carrying a long gun.... I have dirty, ripped clothes.... The battle is going on all around me.... I shoot anything that moves. I really don't want to be here and shoot other people." This led to further recalls that seem to suggest a lifetime as a Civil War soldier.[3]

Such childhood utterances—sometimes even before they have otherwise begun to use complete sentences—about their former families, homes or significant events are fresh and untarnished by the current process of inculcation. However, one must confirm with a second party and document what the child says in order to insure it is consistent with later developments. Unconsciously adjusting memories comes easily.

If possible, the researcher must try to validate the original information before the child has accepted a past-life persona and begins to select his words to reinforce it. Parents' beliefs that it rings true are insufficient validation. When alleged former lives are recent, it is often possible to confirm the veracity of memories by visits with the former family or exposure to situations where the child's alleged memory can be tested. Stevenson and Tucker use this technique.

One intriguing source of self-reporting by children may be emerging from research in transgender studies. Transgender cases involve individuals who "know" their real gender is the opposite of that reflected in their genitals and other features. Many of these cases (probably most, if someone had documented early childhood comments and behaviors) began with statements not unlike hundreds of Stevenson's cases where children's first comments often referred to their "real" identities.

In an anonymous case described in a *Newsweek* article,[4] parents were discussing how to refer to their two-and-a-half-year-old daughter who insisted on wearing boy clothes and playing with masculine toys. "M" overheard them and said, "No, I'm a him. You need to call me him." Shocked, the parents sought advice and decided to let their child behave as he defined himself. It is possible such cases may suggest children have knowledge of an inherited sexual orientation.

Retroactive Hints. Suppressed memories from previous lives appear to unwittingly nudge a person to take actions that appear to have no antecedents in this life. Such actions can only be understood retroactively after the person is aware of a possible past-life connection. The case of Michelle Moshay illustrates this phenomenon.

At age twenty-two, Michelle received a gift from her father who said, "I will pay for a trip to any place in the world you'd like to visit." Instinctively, she chose Tahiti where she visited the Paul Gaugin Museum. The next year she intuitively decided to visit Paris. She didn't meet Peter Teekamp, the likely reincarnation of the soul that also incarnated as Paul Gauguin, until 1991 when she was thirty-five. Six years later Peter thought he saw in her face the image of Mette Gauguin, causing Michelle to recall her youthful, unpredictable trips to Tahiti and Paris.

Self-reported, but verifiable examples (like Michelle's) of possible soul influence that preceded one's consideration of a possible past-life provide useful scientific data. Steve Kern, years later, saw a connection between himself and John Denver in his choice of singing one of John's songs at his senior vocal recital. The self-reported Keene and Kent[5] cases illustrate the value of using such specific data to preclude our natural tendency to see what we look for.

Affinity Clues. The Sorokin/Kee case illustrates an interesting source for clues that may point to a potential case. Known as "affinity cases," the individual feels himself drawn to the life and work of a historical person. After feeling a visceral connection to an aspect of his life, the subject begins to recognize a part of himself in that person.[6] Kee's story began while reading one of Sorokin's books as he started a new phase of self-education.

After graduation from a state college, Kee still felt intellectually incomplete and became a voracious scholar. With the book *Staring Into Chaos* he began an examination of the notion of Western decline through the eyes of Oswald Spengler, Arnold Toynbee, and Pitirim Sorokin. He reported a feeling that he had unlocked some great, hidden secret. He thought, "If everyone could be exposed to this information, the world would be changed."

Sorokin's ideas at Harvard from 1930 to 1955 were the most interesting to him and he read his magnum opus *Social & Cultural Dynamics*. He was struck by the power of its concepts and the degree to which Sorokin's ideas were absent from mainstream sociology. He then read nearly all of Sorokin's works, including his Russian

Revolution autobiography and another he wrote near the end of his life.

At that time, feeling intense dissatisfaction with his career, Kee began a screenplay about Sorokin's life. Sorokin had played an instrumental role in overthrowing the Czarist regime in 1917. He then became a member of the Kerensky provisional government, only to be tossed out when the Bolsheviks took over. Imprisoned, he nearly starved to death under the brutal regime. Pitirim and his wife Elena eventually escaped Russia "by the skin of their teeth." Sorokin then came to America and founded the Department of Sociology at Harvard University in 1930.

A few years after reading Sorokin's works, Kee began debating with authors Neil Strauss and William Howe on their book *The Fourth Turning*. While contrasting the theories of Sorokin with those of Stauss and Howe, other people began to note some of Sorokin's personal biases colored Kee's theories. He initially had a strong reaction to these comments, but then found himself agreeing. This strange, dualistic tug-of-war sensation felt very personal and stuck with him for several months. Then one day, as he described it later, "I began to feel that I *was* Sorokin in a past life."

Increasing Objectivity

The psychoplasm identification process developed during the experiment is designed to collect data from specific, tangible areas of evidence that potentially substantiate reincarnation. It provides a systematic procedure to collect verifiable and comparable data for all the five factors discussed in the last two chapters. Its purpose is to remove as much subjectivity as possible from the discussion of whether two lifetimes are linked by a common psychoplasm.

The reader can access and download copies of procedures and forms for implementing data collection and evaluation from the Reincarnation Experiment web site. The various scales were designed and revised several times over the course of the experiment to improve their reliability, but the reader should keep in mind that it is still a work in progress.

The web-site package contains the full set of five factor scales that define stable components of a human body and its personality.

The hypothesis established in Chapter Nine predicts that, in a likely past-life match, a profile similar to the factor-ratings profile found in today's subject will also be found in the previous incarnation.

The evaluation process estimates a level of probability that the similarities found between a present life and previous life are not based on chance. A high number of correspondences on all factors suggests the reincarnation of the same soul may have been identified. Regardless of the circumstances, in a valid case one should be able to see manifestations of one soul in both lives.

The Integral Model Case Evaluation Forms package available on the web site is self-explanatory. Page one describes the history and rationale of the overall package. Page two provides general instructions for using the various forms. Page three indicates the minimum personal data required for both the subject and previous personality. The next five pages comprise the rating scales used in evaluating the similarities and differences between the two lives.

They are designed to collect evidence relevant to each factor. Each form results in a profile of both the subject and the previous personality for the categories of information it covers. It also helps quantify a level of confidence in the profiles as evidence of a match. The framework on page eleven combines the scores on all five factors into an estimated confidence level, which is a quantitative measure of the strength of a case. The reader may wish to check it out.

Extradimensional Sources

Much of what the public hears about reincarnation comes from nonhuman sources, such as ethereal records or channelled beings. By the time the researcher receives it, it is already second- or third-hand evidence. In the first instance, it is translated by psychics who tap into information from an alleged memory field. In the second, the original source is an advanced being (AB) who is reinterpreted through a trance channel. The ABs come in the form of so-called spirit guides, ascended humans, or other entities who claim to have access to a larger information base on such matters than we humans do.

Secondhand Information. One kind of information is described as being embedded in a universal memory bank (known by some as the

Akashic Records), in an archetypal form or in the individual's energy field. A psychic tapping into these ethereal data banks can obtain information that are interpreted as "past-life readings." They differ from channelled readings in that the psychic does not depend on the cooperation of a nonhuman entity.

The renowned psychic Edgar Cayce (1877–1945) "read" past-life records for thousands of individuals. Archives of the Association for Research and Enlightenment in Virginia Beach, Virginia contains transcripts of these readings. Many of those who received the readings felt resonance with the material, but in most instances they had no direct way to confirm it.

A large number of present-day seers and mediums claim to tap into the Akashic Records or other sources for individuals interested in their own reincarnation history. The readings offer clients fragments about previous lives, often reinforcing hints taken from the client's behavior, appearance, or comments. These exchanges often insinuate personality or karmic issues, but usually provide little actionable information.

Thirdhand Data. In this situation, the original source is two steps removed from the subject or researcher. Marcia Schafer, a traveller well-versed in extradimensional communications, counsels caution about such sources. In her *Confessions of an Intergalactic Anthropologist* she writes, "Most people aren't aware that translation errors can happen.... the (channel) is a filter through which all information borrows his or her color of bias and knowledge." [7]

With both the AB and the psychic involved, the researcher has information from a source that can only be accessed by a channel who cannot confirm its veracity. Given the translation problems highlighted by Schafer, the chances of glitches in communication are high. In spite of these obstacles, experience has shown some very valuable clues can come from such sources.

The Swidersky/Alexander case provides a good example of an AB providing clues to a subject researching his own case. In 2000, Ken had reluctantly accepted reincarnation as a likely explanation for many of the anomalies in his life, but he had nothing specific. At that point he began to consider a 1981 dream as possible evidence of a particular past-life. After meeting Walter Semkiw and Kevin Ryerson, he decided in late 2005 to contact the AB Ahtun Re through Kevin. Ahtun Re said he could not identify a specific past life, but gave Ken several clues.

Ahtun Re informed Ken one previous personality had been involved in the technical development of the U.S. Navy airship Akron as both a civilian and a member of the military. His speciality was related to the design of the steering mechanism. The alleged personality had lived in Akron, Ohio where the craft was built. Ahtun Re further said the man had lived near and worked with Ken's grandfather in Akron at that time. In searching for the name Ahtun Re had only been able to come up with "Hackman."

With these specific clues to focus his personal research, Ken was able to verify that the airship had been built in Akron, Ohio by a U.S.-German partnership (Goodyear-Zeppelin). It had crashed in April 1933 off the New Jersey coast, killing all but three of the seventy-six officers and seamen. While the roster had no one named Hackman, only one causality was listed as being from Akron, Ohio. This was enough for Ken to uncover the story of Tony Swidersky who was born in Pennsylvania in 1902.

More details are found in later chapters. Ken's diligent work has resulted in him having a solid case for the likelihood of a linear past-life match with Tony. In addition to understanding himself better, Ken's journey of self-discovery has put him in contact with James, Tony's surviving son.

Cases like this one illustrate how information from nonhuman sources can be tangible enough for the individuals involved to develop corroborating evidence from primary sources. While claims from third- or secondhand nonhuman or human sources cannot be used as the basis for a verifiable past-life identification, they may still be very helpful. The point to be made here is the need for objective verification before acting on the proposed link.[8]

Confirming Nonhuman Sources. Uncontrolled use of extradimensional sources or human psychic perceptions poses challenges to scientific validation. Many people assume that AB or spiritual sources are divine beings and must, by definition, know everything. Humans forget that the knowledge of all conscious entities is limited by nature to the scope of their senses and their capacity to accurately understand what they can perceive. When evaluating the value of AB-sourced information one should address several issues:

The record of the source in providing information that is consistent with scientific tests. The specificity of the information and its reliability when tested by other means. The possible agenda of the nonhuman being in providing the information. And finally, any vested interest the human channel has in the outcome of the research.

On the last point, a psychic who wants to please his client may subconsciously guide the inner translation process toward what he thinks the client wants. Channels who are also involved in research on reincarnation will have developed biases on the topic and may also have a priori opinions about a specific case. Remember Schafer's caution about "translation" mentioned above.

The methodology used by Gary Schwartz in his groundbreaking book *The Afterlife Experiments* may be helpful in avoiding vested interests. In testing extradimensional messages to humans from nonphysical beings, he used several psychics in a double-blind experiment in which each psychic had no knowledge of the person for whom the otherworldly message was intended.

The experiment described in this book has shown that careful handling of data from nonhuman sources can often lead to evidence that will satisfy the skeptical but open mind. Good detectives never automatically squelch relevant information regardless of the source. They review all the calls purporting to have leads that may solve the crime, but they validate them before they make an arrest. If professional reincarnation researchers act with no less integrity, they will conduct a thorough investigation of different types of evidence before claiming a past-life match.

Increasing Reliability

An increase in the reliability and validity of data on which to base proposed past-life matches can be achieved by attention to three areas: sources, types of evidence, and evaluation. The above discussion of human and nonhuman sources makes it clear that, as in good intelligence or detective work, the researcher must corroborate information from a single source by an entirely different source and method. In other words, two dreams or two psychics with the same data do not demonstrate corroboration. Intuitive material must be corroborated with tangible evidence. And, several types of tangible evidence are required to make the case persuasive.

Using a variety of sources and data is necessary to demonstrate that the evidence for a specific reincarnation cannot be reasonably attributed to chance. The Integral Model, with its use of tangible data from a variety of sources, undercuts the randomness argument and enhances the probability argument.

The strength of this model lies in its incorporation of interlocking factors that represent the core traits of a real personality. One can start with any factor with the expectation that in a valid case high correspondences in two or more areas will be accompanied by high correspondences in other areas. If one finds an apparent facial and body-type match, he then looks for mental and emotional correspondences. If equally strong matches can be found, the researcher then looks at career choices and life patterns.

Regardless of where one begins, if a case is valid, its interwoven threads should lead from one to the other. A false past-life identification very likely lacks this internal reliability.

With regards to the evaluation process, multiple evaluators should be involved. Even when using objective rating scales, care must be taken to avoid rater bias. In the early stages of this experiment, I realized that I and others involved already had opinions about the strengths of various cases. For that reason, ratings were based on substantiation in the form of physical measurements, documents, third-person reports, or behavioral evidence.

Regarding physical measurements, the biometric scoring system used is still in its infancy and subject to variations caused by lack of precise data points. In this situation, different people are asked to duplicate measurements. When significant differences result, an average of them is used for comparison. Even with these precautions, the measures can be considered only proximate indications of the actual variances.

On the rating scales, to protect against potential rater bias, each subject can offer specific data that might affect his or her personality-factor evaluation. To improve objectivity, the person developing the subject's ratings should not be the same person who does the ratings on the previous life. This keeps preconceptions from shading the scoring in the desired direction.

A Team Approach. Research should be a team effort to compensate for missteps in one or another team member's approach. A hypnotist may unwittingly use leading questions or structure a session to point towards a desired outcome. The researcher may inadvertently provide a psychic or channel with information that colors later exchanges. Some member may realize he has had vested interest in the outcome.

Contaminated evidence can come from the subject as well as the therapist or researcher. If an individual has a strongly vested interest in a particular past-life identification, his meditations may lead him to imagine supporting material. Or, he may subconsciously program his dreams to fulfill his wishes. For this reason, using personal material generated after the subject has become convinced he is the reincarnation of a particular person may be misleading. It is best to use only data dated prior to the possibility of self-programming.

The polygraph is of no value in establishing the veracity of a subject's or psychic's data. The so-called lie detector test registers physical reactions that reflect whether the person truly believes what he is saying. Believing his statement is the truth can obtain a positive score from the polygraph examiner, but that does not mean it is true. The same problem exists with the so-called process of "muscle testing." If a person truly believes it, the body confirms it.

Most of the cases in the experiment began as self-identified initiatives. James-II had intuitively postulated a past-life connection to James more than two decades before he became involved in the study. Keene began his search after an emotional breakthrough on a battle field visit. Teekamp and Moshay were stimulated to study their lives due to the comments of others. Kee's self-start is described. Laird's case was self-initiated, as was Karlen's. Alexander started with dreams, but was encouraged by clues from an extradimensional source. Others were stimulated by dreams or the comments of others who thought they recognized a historical personality in them.

The point to remember is that regardless of how a subject was stimulated to begin the search for a connection to the past, each case has been evaluated solely on the basis of evidence collected in the manner recommended in this chapter. All extradimensional information has been treated as no more than clues to confirmable evidence.

Caveat Lector! The reader should beware that this experiment has not always been able to observe all the standards it promotes. Some of them grew out of the need to test different methods in its early stages. This made it difficult to avoid cumulative biases as the process evolved. I believe that, despite these shortcomings, the model speaks for itself through the evidence described in the coming chapters. I invite others to replicate the experiment in a more formal and controlled manner. My research is ongoing and further advances will be published on the Reincarnation Experiment web site as they become available.

You can access and download the Integral Reincarnation Model Case rating scales and evaluation forms used in the experiment to begin your own self-directed research. Various forms of feedback, including biometric comparisons of your own visual images along with those you have on the possible previous life, may be obtained from people involved in the experiment.

Research Your Own Past-Life Case. The Integral Model can be equally helpful to the professional researcher and the motivated individual. The forms on the web site make it possible to establish and evaluate a potential case. Engaging in the data-gathering and evaluation process can offer insight into old behavior patterns in the present incarnation and help sort out today's dilemmas. One can go through the assessment of the present life without any idea of what the most recent incarnation might have been. Simply engaging in the project may awaken past-life memories or reveal an existing affinity for a historical personality.

Part III

EVALUATING EVIDENCE

A sample of robust cases is used in this section to demonstrate how the model and the process work. For each of the factors (categories of evidence), this section examines the value of the linear reincarnation hypothesis as an explanation for correspondences between our subjects and their proposed past lives. Chapter Thirteen starts with the physical evidence: facial geometry and other body features. Chapter Fourteen covers the evidence for the reincarnation of cognitive traits.

Chapters Fifteen through Seventeen cover examples of the emotional, interpersonal, and creative factors that lie at the core of the personality-based model of reincarnation. Each of the featured cases is illustrated in terms of both support for the hypothesis and implications for the individuals involved.

Chapter Eighteen demonstrates the many ways that subconscious memories from previous lives may influence the seemingly inconsequential decisions people make everyday. Chapter Nineteen explores the possibility that not only are our personal lives affected by reincarnation, but that many of our significant relationships may also have roots in previous incarnations.

Chapter Thirteen

Physical Phenotypes

The reader may wish to start this chapter by reviewing the front cover of this book and the James/James-II comparative triptych at Figure 4. The resemblance between two men living almost two centuries apart—on different continents, but in similar romantic circumstances—is uncanny, to say the least. The variances between their facial geometry ratios are what one might expect from a good artist painting portraits of the same person a few months apart. However, facial structures are only part of the genotype that the psychoplasm apparently carries forward from life to life.

Chapter Ten explained how biometric science points to the genetically-based physical characteristics that may be most relevant for reincarnation research. The reader may recall that the most genetically stable features are facial geometry, ear forms, hand and finger proportions, voice, odor, body types, and special birthmarks or scars. However, we can only measure a phenotype. Why it sometimes varies from the genotype was discussed in the previous chapter.

Pictures and descriptions in this chapter show why I believe that the psychoplasm duplicates certain genotype features from one incarnation to another. In order to go beyond one's subjective perception of these images, they can be measured to determine how closely the two bodies from different lifetimes match. The variances between the two set of images can then be compared to randomly selected, unrelated individuals to discount the the role of chance.

With a limited data base, we cannot yet establish precise levels of variance for different degrees of past-life relationships, or even all present-day relationships. It tentatively appears possible to assign at least three general ranges of scores that suggest different levels of confidence. The close matches found in very robust cases indicates some confidence in the process. I believe the methodology can and will be improved as technology and financial resources become available.

As you review the sets of portraits and pictures, remember the two parties are not biologically related to one another and their lives do not overlap in time. Yet, in spite of such gaps in the DNA chain, many subjects appear to be identical twins of their alleged previous incarnations. This is precisely what the reincarnation hypothesis predicts when it posits that legacy-genotype traits carry forward to the next birth.

In the strongest cases, all physical features about which we have information from the previous-life can also be found in the present subject. This suggests two things. First, researchers should collect data on all relevant physical features. Second, they should not ignore significant gaps in phenotype correspondences between the two lifetimes. Scrutiny of such gaps is required to avoid a false past-life identification. Several cases now illustrate the total comparison process.

Facial Geometry

Generally, good photographs are available on present-day subjects. For previous personalities who lived during the last century and a half, it is sometimes possible to obtain photographs, but they are usually not posed or precise enough to match the accuracy required by photo-ID technology. This occurred in the Tony/Ken case. The one, slightly unfocused, photo of Tony available during the experiment does not give us the appropriate profile.

In dealing with historical personalities, with luck, high quality, hand-painted portraits may be available to match against modern photographs. Facial geometry measurements from portraits done by different well-known artists are surprisingly consistent. However, when there are concerns that old portraits may not accurately reflect reality, one can use the averages of ratios from two or more portraits to compare with the ratios on a modern photograph

In the James/James-II composite in Figure 11, involving a nineteenth-century portrait and a twenty-first-century photograph, the average variance is about 2%. That caliper-measured variance ratio is close to the minimum required by ID experts using a live person and a powerful digital camera and computer.

James and James-II Aging Alike?

Courtesy of Walter Semkiw   Fig. 11

While there are a number of contemporary portraits of Dolley, only two, not-very-good photographs were made late in her life. The best measurements from both media are used here. The comparison of several such measurements between Dolley and Kelly's full-face photograph results in a variance of less than 3%. This variance level indicates strong correspondences between the geometry of the two faces. The variance between Charlotte's full-face photograph and the same images of Dolley is about 7%. This variance level may also indicate family resemblances. (See Figure 12, next page.)

The Sorokin/Kee triptych at Figure 13 shows a very strong resemblance, but the lack of good full-face photographs from Pitirim's life comparable to Lorin's current age made precise comparisons difficult. The average of four efforts to compare an older Sorokin photo with youthful Kee photos yields measurements in the 5% range of variance. That percentage includes a phenotypic-nose variance that usually does not reflect the underlying bone structure. This means the genotype variance is probably smaller.

The four photographs of John and Steve at Figure 7 illustrate the look-alike contest problem noted on page 85. When one glances at the top photos, the immediate response is "They look alike!" This results from several factors. If one is already expecting them to look like twins, that conclusion is easily made. Hair styles, glasses, guitar, and clothes suggest a match. To avoid such subjective projections, the less adorned photos on the bottom must be analyzed.

Three Women Separated by Time and Space with Apparent Soul Connections.

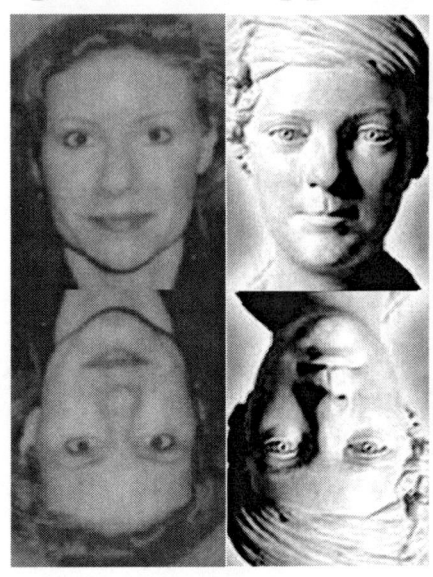

Photograph of mature Charlotte with youthful sculpture of Dolley. Note interesting comparison with mixed media and inverted images.

Facial features in the younger and older images of Dolley and Kelly yield useful comparisons. Kelly's glasses symbolize the life-long poor eyesight of all three women.

Charlotte 	Dolley 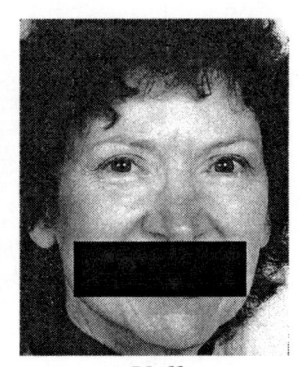	Kelly

Fig. 12

Two men, different ages in two historical eras. One genotype?

Pitirim Sorokin   Lorin Kee   One genotype?

Fig. 13

The faces reveal differences in facial geometry: One is long and angular, the other more oval and softer. Use of the six measures to obtain three ratios takes into account the different sizes and positions of the photographs. According to those measures, there is a variance of 12% in their eye sockets and underlying bone structure. The difference between their nose architecture is 28%. In their facial proportions the difference is only 2%. This gives an overall 14% variance.

While the John/Steve variance of 14% may be smaller than between some random persons, it is much larger than the robust past-life cases. At this point, it is difficult to predict that the two heads came from the same psychoplasm. It may be in the range of a soul birth-family connection.

Body Types

The apparent genotype transmission by the psychoplasm from an earlier life involves more than facial features and birthmarks. This is indicated by the fact that historical cases that include full-length images reveal body-type matches with the subjects. Even with only descriptive accounts, when no photographs or portraits are available, body-type comparisons are possible.

When comparing body types, one must be careful to take into

account phenotypes changes caused by diet, health and other factors (discussed on page 91). To preclude such changes, comparisons should be made as early in life as possible. Note should be taken of circumstances relevant to phenotype changes when evaluating the differences.

For instance, one source described young Dolley as a "slight figure - about 5'6," [with] delicate oval face..." Later, another mentioned her "heavy eyebrows and long lashes, black curly hair and brilliant skin. ... with a handsome bosom and erect carriage." By age forty-one, Dolley was described as a "portly, buxom dame."

As young women, Charlotte and Kelly, with fair skin, feminine ectomorph forms, and heights about 5'6," mirrored the body type that portraits and written accounts suggest was exhibited by young Dolley. With different diets and lifestyles, neither Charlotte nor Kelly produced over time the middle-aged body Dolley reportedly developed.

In depending on historical reports, one must also keep in mind the norms of that era and the subjective perspectives of the reporters. The literature reveals that people who liked James and Dolley described them larger than those who were not their friends.

A lifestyle effect can be seen in the Keene/Gordon case. Both men were six feet tall. While Keene's cessation of smoking resulted in a weight of 200 pounds in later life, they both were trim and muscular in their younger years. Their body postures and arm-folding patterns give the appearance of the same bodies at work (see Figure 14). Photos reveal that Kee and Sorokin were both tall, lean ectomorphs. Photos of Ken and Tony reveal well-built, muscular mesomorphs.

Figure 15 is a photograph, taken during a research meeting, of men who may be the reincarnations of John Adams, James Madison, and Thomas Jefferson. The man on the left recalls the eighteenth-century images of Adams' stout endomorphic profile. The center figure resembles the tall, muscular mesomoph body of Jefferson. The short ectomorphic body type could fit Madison's short, slight build. The basic types are consistent between the two time periods, with small phenotype changes. If their Revolutionary-Era contemporaries could see this picture, it is likely that they would believe the three men to be John, James, and Thomas in stage costumes.

Confederate General and Connecticut Yankee

John B. Gordon   Jeffrey J. Keene

Note similar facial features with slight phenotype differences above and comparable body types below. Adult weight gain by Keene results in only superficial variances.

Fig. 14

Three Body Types

The body types are ectomorph, mesomorph, and endomorph. Men photographed at a research meeting reflect the types of John Adams, Thomas Jefferson, and James Madison.

Man on the left mirrors the endomorphic type of Adams. Center figure resembles the tall, muscular mesomorph body of Jefferson. Short ectomorphic type, discounting the phenotype waistline, could fit Madison.

If contemporaries of U.S. Presidents numbers 2, 3, and 4 saw them, would they perceive these men to be John, Thomas, and James in weird stage costumes?

Fig. 15

Ear Forms

The form of the ear is considered by biometricians to be among the most enduring genotype features. The profile pictures of Keene and Gordon on pages 151 and 152 in Keene's book *Someone Else's Yesterday* provides clear evidence of this feature. Photographs of James and James-II in several examples also illustrate the correspondence of ear forms. Note the overlay of James's exposed right ear on the partial ear of James-II in the triptych. Positioning and size of the ears of both men appear to be close to identical.

With only one—slightly unfocused—photograph of Swidersky, measurements can be made of his and Alexander's ear. Their ear forms appear to be similar and the ratios of ear-length to head-height vary by only one percent. Using not very clear Sorokin photos, it appears he and Kee have very similar ear patterns and their ear-to-head ratios are the same. In addition, the ratio of Kee's ear length to lower-face length varies less than three percent from Sorokin's.

Comparisons of the ear forms of historical women are not as easy to make when only formal portraits are available. In a few youthful portraits, it appears that Dolley's ear proportions are somewhat larger than average, but her ears are not visible enough to make good comparisons with Charlotte's and Kelly's ears.

Hand/Finger Proportions and Voice Tone

Biometric data suggests these two physical features are also genetically stable. The photos in Figure 16 show the similarity of hand and finger proportions between Marilyn and Sherrie. Similar measurements have been made for James and James-II. During the experiment, an interesting comparison was made between the hands of James-II and George Mettler. Ectomorph James-II had wider hands and longer fingers than tall and broad George's hands.

Matching Hand Proportions

Marilyn Monroe   Sherrie Laird

Fig. 16

The size of George's hands are inconsistent with his overall bodytype. As a basketball and football star in high school and college, he was frequently asked, "How did you hold that ball and throw that pass with those hands?" In a comparison of Mettler and Johnson phenotypes, this anomaly is of interest. Two popular eighteenth-century portraits of the large and powerfully-built Samuel Johnson show he, too, had especially small hands for his body. As comparative evidence for a past-life match, two irregularities result in a positive score.

James was known to "not have a strong speaking voice."[1] James-II is burdened with the same inadequacy. His voice contradicted at an early age his desire to be known as a great public speaker. He decided, like James before him, that his pen was a more effective medium for his communications than his tongue.

Mette Gad was described as having a gruff voice and Michelle Moshay has the same trait. And, they both have a similar habit. Both smoked (cigars for Mette) which may account for the voice, but Michelle (who no longer smokes) says her voice has been "raspy" since childhood. It may have been influenced by drinking an inadvertent dose of a strong medicine, but regardless of cause, the two voices do correspond in a phenotype comparison..

In most of the cases in our experiment, no voice recordings exist for the hypothesized earlier incarnations. An exception is Marilyn Monroe with her innumerable voice samples. Finkelstein's attempt (as reported in his book) to get a credible laboratory comparison of Marilyn's and Sherrie's voices proved fruitless. I hope to obtain tape-recorded lectures or speeches by Sorokin and samples of voice recordings in other cases. Reports of the results of any voice analyses will be reported on the Reincarnation Experiment web site.

Other Genetic Effects. A genotype problem shared by Dolley, Charlotte, and Kelly is one of poor eyesight. Dolley's difficulty seeing was described as "excessive at age twenty-six." Kelly and Charlotte have depended on glasses all their lives. Another problem of note is Dolley's 1805 bout with an ulcerated knee that posed the danger of amputation. Kelly suffered with a hip problem that required modern hip-replacement surgery. Both required similar care from their respective husbands while they were preoccupied with writing and editing duties.

Considered "sickly" as youths, James suffered with a "nervous disorder" and "bilious attacks" and James-II suffered with "bronchial asthma." In the James-II case, his health problems were aggravated in childhood by lung damage from flames when his clothing caught fire. James reportedly had a very sensitive nervous system and complained of an "unsettled" and "feeble state of health" through most of his life. James-II suffered early from sensitivities of the stomach and skin. James suffered from rheumatoid arthritis that became debilitating in his later years. James-II uses dietary supplements to deal with the same issue.

Special Markings

Many of the University of Virginia cases discussed in this book focused on unique body markings present from the time of birth. Tucker, in *Life Before Life*, highlighted the case of John McConnell, the slain New York policeman who apparently reincarnated as his daughter's son William. Young William has birth defects located in the same areas where his grandfather suffered his fatal wounds. Other trauma-related marks were discussed earlier, but not all such correspondences are trauma related.

Laird describes becoming subconsciously aware of a mark linking her to Monroe. "When I was eleven or twelve, I was sitting on my aunt's knee and we were talking about the beauty mark just above my lip [where one appeared on Monroe]. I asked what it was. It looked stupid to me. She began singing [a Monroe song]..... The lyrics just echoed in my head, as if coming down a long corridor of sound from the past."[2]

Self-inflicted changes to the body can also have the effect of making the subject's physical appearance more like its reputed previous life. Jefferey Keene accidentally ran into the gym wall in junior high school where his encounter with a nail resulted in a scar over his left eye. During his adult research on a past-life connection with John Gordon, he learned that his nail scar matched a bullet wound scar over Gordon's left eye.

Michelle Moshay, whose genotype matches that of Mette Gad-Gauguin, describes how her body assumed the appearance of a woman who could have given birth to five children (as Mette did in the eighteenth century). A C-section-like surgery to remove an ovarian cyst left Michelle with the scars and slack muscles in a protruding abdomen similar Mette's older, maternal body.

Peter Teekamp has a similar tale of the body "catching up with its past." Paul Gauguin broke an ankle which did not heal properly, leaving a twisted limb. Peter was not born with a misshapen ankle, but when he accidentally broke his elbow it was cast incorrectly. His elbow now looks like a misshapen foot. These examples and others suggest that some aspect of a subject's consciousness may try to complete the replication of a previous body map.

The case of Brian O'Leary may indicate how the energy of the psyche re-manifests, even if only temporarily, markings from a past-life, one's own or another's. In 1998 in Glastonbury, England for a conference, he had a spontaneous vision of being in shackles. During a 1988 visit and the 1998 one, he stayed in, and next-door to, a room in the inn that was reported to be haunted. After the vision, he broke out with rashes in exactly the areas where the person in the vision was shackled: lower arms and legs, crotch, and neck. Brian's wife confirmed this report.

If the subconscious can cause Brian to break out in rashes, it is reasonable to assume that subconscious influences could result in physical effects like those mentioned earlier. The effects could be produced psychosomatically or through the stimulation of individuals to take actions to produce the mimicking effect. A stable psychoplasm (soul genome) could provide the continuity.

Chapter Fourteen

Cognitive Cerebrotypes

The histories of Dolley and Kelly show they share similar intellectual interests, to the point of employing the same media to satisfy them. Despite different societal opportunities for women's education and work outside the home, the record shows what seems to be the same mind at work.

Dolley often expressed poetic aspirations. For instance, she wrote to James on 5 December 1826, "... in truth, I am too busy a House keeper to become a poetess in my solitude." She sent verses to friends with tender sentiments and moral truisms. In letters she reached for poetic expression: "Imagine if you can, a greater trial of patience than seeing the destruction of a radiant patch of green peas, by frost!" She shared her poetic drafts with some. Her sister Lucy wrote, "... the Verses you sent were very good indeed and caused us a hearty laugh and me a *kiss extra*."

Dolley tried to keep up with new novels and pleaded with her correspondents to send them her way. "Bye the bye—do you never get hold of a clever novel, new or old that you could lend me?" she asked in a letter to her niece. Dolley practiced her French on France's diplomats in Washington. Kelly studied French in college and used it for her graduate research.

Kelly shares Dolley's interest in poems and novels, organizing poetry and book events for friends to share and critique their work and reading. She has realized Dolley's aspiration by becoming a prize-winning poet. Like Dolley, one of her favorite themes is nature, focused on the seasons and the beauty of her gardens. To insure that she keeps up with current fiction, Kelly subscribes to literary magazines and keeps two or three novels in progress at all times.

Though Dolley had only a basic Quaker-school education, she steadily developed her reading and writing skills throughout life. Her conversation was as lucid as that of most other upper-class Americans of that era. In the second half of the twentieth century, Kelly could take her formal education to the Ph.D. level with research in the eighteenth and nineteenth-century English poetry and fiction that Dolley read and enjoyed. She teaches at the university level and writes professionally about the subjects Dolley discussed with her circle of friends.

Dolley showed a keen interest in international as well as domestic politics. Deferring to the times, she demurred from public political discourse. But, men and women highly respected her private views on the subject. Kelly also developed a wide knowledge of domestic and political issues; she chaired international seminars with insight and aplomb.

Charlotte equally shows an inquisitive mind. In her early grades she often received the highest grade in her class. With youthful interest in political issues, she received her bachelor's degree in political science. Pursuing a lifelong program of self-education, she continues a wide range of reading. After an interest in novels in her younger years, her adult reading tends to nonfiction. Her areas of study are ancient history (including speculative history), modern history, metaphysics, the paranormal, religions, and, with a special interest, health and nutrition.

Charlotte challenges traditional religious beliefs, embracing new metaphysical ideas and alternative views of reality. She leads occasional public discussions in some of these areas and responds quickly and intuitively to new questions or issues. Her written treatises for limited distribution are on the topics of diet, health, and reincarnation. She is a dedicated user of e-mail and information packets to communicate with a wide circle of correspondents.

Dolley and Kelly are devoted to letter writing. Dolley personally answered every condolence letter (many from abroad) she received after James's death. Kelly never fails to write gracious thank-you notes on every occasion covered by traditional protocol. Each makes sure friends know they are cared for, maintaining faithful contact with a wide circle of correspondents.

Dolley and Kelly each highly valued the development of her son's mind. In her will of 13 May 1794, Dolley wrote, "... the education of my son is to him and to me the most interesting of all earthly concerns, far more important to his happiness and eminence in life than the increase of his estate..." She and James sent John Payne to a private school at age thirteen. They sent him abroad for further study. Kelly enrolled her son in a private college in Hanover County (Dolley's youthful home) and added travel and internships to expand his experience.

Cerebrotype Traits. If the visible, physical evidence of the genotype carry-over seems persuasive, cognitive correspondences may be even more compelling. Science has made much progress in linking physical features to simple inherited DNA patterns, and perhaps associated, bioenergetic fields. But, how do the minds of an Albert Einstein and a Thomas Jefferson with all their mental quirks and levels of complexity get transferred to the next incarnation?

Chapter Eleven set forth the hypothesis that mental as well as physical genomic material in a psycho-energetic field links one incarnation to another. While there is yet no direct validation of the psychoplasm concept, this chapter's array of cognitive correspondences between subjects and their alleged previous lives calls for an explanation like it. The above descriptions of Dolley, Kelly, and Charlotte demonstrate some of their respective cognitive traits.

I believe stable underlying traits can be identified from personal archives—the records and commentary most of us produce in the course of daily living. This view is based in my research with David McClelland at Harvard (reviewed in Chapter Seven) that identified basic needs or motives by analyzing the mundane detail in personal documents. This chapter illustrates how modes of thought can be seen in people's fiction, nonfiction, poetry, letters, and e-mail.

Modes of Thinking. For James's cognitive evaluation, I begin with renowned Presidential scholar Ralph Ketcham for comments. In his highly praised book *James Madison: A Biography*, Ketcham uses the phrases "keen minded and brilliant" [with an] "intricate quality" to describe James's mind as seen by his contemporaries. Using third-person accounts made during James's life, he concludes that James, "by scholarly and intellectual habit... weighed matters carefully, sought subtle and sophisticated insights, and suspended judgments as long as possible..."[1]

To give a European perspective, Robert Allen Rutland, another Madison scholar, adds to this picture with similar assessments. He quotes a French diplomat who considered James "well-educated, wise, temperate, gentle, and studious." Jefferson said James's contribution to the *The Federalist* was "the best commentary on the principles of government which was ever written."[2]

On James's approach to analysis and development of a thesis, Ketcham quoted the terms applied by a friend of James during his later years: "... ample research ... full, clear and deliberate disquisition ..." This friend, Hugh Blair Grigsby, had the impression that James "regarded himself rather as the representative of the truth than the exponent of the doctrines of a party or even of a nation ..." Ketcham further characterized James as scholarly, an abstract thinker, having a bookish inclination, being a lover of learning, deliberate and steady, and possessing a shrewd realism.

To get a third-party perspective on James-II, I use material from some of his U.S. government performance reports for comparable perceptions of his mental aptitude and style. One described him as "unusually alert, intelligent, and well educated ... energetic, personable, and eager to work." It further commented that "his written work is well-organized [and] his political reports have been pertinent, concise, and perspicacious." Another said he was "liberal, humane in outlook, and, thus, totally free of bias." He was described as "highly articulate in stating a case."

More formal measures of one's mental attributes are available today than in James's time. The alleged reincarnated soul's cerebrotype capacities in James-II have been officially recognized through his membership in the Phi Beta Kappa Society (PBK) and high IQ-based Mensa. PBK, the first of the Greek-letter fraternities in America and an honorary society for the top students in liberal arts colleges, was founded in December 1776 at William and Mary College in Williamsburg (the colonial capital of Virginia). As a footnote to history, had James not rejected W & M in 1769 in favor of Princeton, he might have helped found Phi Beta Kappa.

Writing, a form of thinking, is clearly a manifestation of the soul's cognitive software package. Professional linguistic analysts assess pairs of communications contained in written documents, audio or video tapes, or other records to indicate a level of probability that they are from the same person. Intelligence agencies and police departments make use of their services for obvious reasons. I believe these methods can also assess the reliability of reincarnation matches.

The factors used by these analysts include the complexity of thought reflected in use of compound sentences and the number of words per sentence. Use of preposed or adverbial clauses is another standard. So is starting sentences with phrases such as, "There was no way to escape." Patterns of active versus passive voice differentiate speakers and writers. These and other aspects of syntax (style of arranging words), morphology (word formation and compounding), and semantics (use of words for effect) seem to be part of the soul legacy.

James's writing style was known for its complex syntax and logic. One had to pay close attention to his twists and turns to understand him. An example of his convoluted sentences can be seen in the *Federalist # 37*:

"As our situation is universally admitted to be peculiarly critical, and to require indispensably, that something should be done for our relief, the predetermined patron of what has been actually done, may have taken his bias from the weight of these considerations, as well as from the considerations of a sinister nature. The predetermined adversary on the other hand, can have been governed by no venial motive whatever. The intentions of the first may be upright, as they may on the contrary be culpable. The views of the last cannot be upright, and must be culpable."

Judge for yourself whether James-II reflects the same writing style used by James. The next paragraph, written about 1980, exemplifies James-II when writing without an editor:

"During the last two hundred years decisions have been taken, consciously or by default, that moved the operation of the American national government away from these novel principles and in the direction of more traditional social institutions that lack the built-in renewal mechanism. That form of bureaucracy, nurtured by a symbiotic relationship with Congress, partisan politics, and a large range of special interests and supported by myths apparently immune from serious intellectual challenge, has now fallen into the same rut that spelled decline of all its predecessors."[3]

In an undated, hand-written draft of a potential cosmological treatise titled as "Symmetry of Nature," James revealed his layman's affinity for the natural philosophy perspective of the scientists of the period. The introductory paragraphs read:

"The planetary system, the greatest portion of the universe, as yet brought under human observation, is regulated by fixed laws, and presents most demonstrably a scene of order and proportion. From analogy we conclude that the whole universe, if it were equally understood, would exhibit equal proofs of a like arrangement. The general aspect of the earth leads us to much the same plan of nature. Order and symmetry equally appear in the great outline and in the most minute features of it."

"In the interior structure of the earth and in the mineral kingdom, which lies chiefly below the surface, less regularity is indeed to be perceived. But even here nature has her laws, and if they are not more known, it may rationally be ascribed to the imperfect insight to which her work is subjected. It is extremely probable that if the whole earth could be laid open & thoroughly examined from its center to its surface, in every direction, it would not only exhibit proofs of a general plan, but that many of the subterraneous parts, which on examination have appeared to be thrown together without order or design, would be found related to other parts now unknown, or to the whole, so as to give meaning & method to what has at present no visible trace of either."

In a similar burst of amateur scientific enthusiasm, revealing his avocational reading of historical and modern discussions of issues facing science, James-II wrote:

"... we can apply insights from the microcosm to other aspects of life in our universe. An example of the correspondence between levels is the behavior of atoms roughly simulating the actions of heavenly bodies. In an experiment with the "Rydberg atom" in which electrons are artificially placed far away from the nucleus, the orbiting electrons behave like planets. Such examples do not definitively prove the Principle of Correspondence, but they do provide the basis for such a hypothesis."[4]

"... The Principle of Correspondence... means that one can infer the nature of smaller-scale entities from the characteristics of larger, more distant realms, and vice versa. The dynamics of cells parallel those of galaxies. Just as a small laboratory or computer program can simulate the behavior of stars billions of light-years away, the local consciousness of an individual can confer with the universal ultimate consciousness that existed when there was only the Word."⁵

Separated by almost two centuries, the two passages suggest James and James-II have not only a similar level of intellect and knowledge, but the same style of reasoning. Both use analogy and inference to extrapolate from the known to the unknown. On the cerebrotype rating scale, James and James-II match on almost all variables. James-II appears slightly more intuitive and freethinking, but that may be because we have less of a public record of James's informal thinking.

This example of parallels in the writings of what appears to be one soul in two different incarnations may also provide insight into the dynamics of reincarnation. First of all, James and James-II share a belief in the symmetry and order of the universe and arrive at a common approach to interpreting evidence from nature itself. Their common areas of interest are self-evident. Both are intrigued by the majesty of the universe and wish to understand the laws of nature.

Where James apparently reflected the insights of Greek and Renaissance scientists and Enlightenment writers, James-II was able to validate them in a modern laboratory. Given recent discoveries that extend our vision backward into history and out into the universe, James-II is able to link his and James's views to a historical source (the Egyptian Toth) and to cutting edge physics (the Rydberg atom).

Perhaps of greater importance for the concept of reincarnation, these two passages suggest a progression of the psychoplasm's access to knowledge about the universe. As the scope of individual conscious awareness increases, the insights seem to become more refined and grounded in the soul's legacy to the next incarnation.

For further evidence of how the cerebrotype carry-forward seems to work, I turn to the Sorokin/Kee case. Sorokin's last phase of writing began to examine metaphysical concepts such as altruism and "supraconsciousness." Interests in this area came very early in Kee's life (perhaps suggesting the soul picked up where it left off in the previous life). Natural spirituality and the physiology of enlightenment have been Kee's primary interests for some time. He, taking advantage of leaps in technology, started a web site (www.the gnosticoracle.com) for public discussion of such matters.

Both Sorokin and Kee share a similar fascination with determining where society is headed. Kee, like Sorokin, believes that our culture will soon face an upheaval that will ultimately lead to a more harmonious society. He feels that his purpose for this incarnation may have something to do with participating in this cultural transition. Let's compare a view of Sorokin expressed by youthful sociologists a year after his death with some of Kee's Internet postings.

According to a 12 August 2002 biographical overview by the Dead Sociologists' Society, as a conservative libertarian and a Christian anarchist, Sorokin never lost his peasant distrust of the centralizing state, a distrust that was reinforced by his experiences during the Russian Revolution. Thus, Sorokin had little in common with his American counterparts.

"The 1969 American Sociological Association meeting brought recognition of a different type.... wearing the button 'Sorokin Lives!,' radical students rebelled against a corrupt power structure, the Vietnam War, and a servile, impotent sociology.... In (him) they glimpsed a sociologist who understood and fought against human suffering. His was a strong, angry voice, opposed to brutalities of the modern age and committed to a science of reconstruction... (He) was not only relevant, he was essential, and on that day they celebrated the sociology of the pariah."[6]

From the skeptical acts of Kee, it appears the same fired-up, critical mind lives on.

In a public petition challenging the power of the privately owned Federal Reserve Bank, Kee wrote, "The Central banks will destroy us all—including you. Crush the beast while there's still time." Referring to his own web site he writes that it:

"... is meant to be a "jumping-off point" for each individual to begin his/her own research. It was created simply to sound a warning... The message is this: The United States is hopelessly bankrupt, the Federal Reserve is a massive scam and the day is fast approaching when Americans will wake up to discover that their only real wealth are the clothes on their backs and the food in their pantries. The elite in our country are quietly and methodically making their own preparations for what is to come. Hopefully readers of this site will "follow their gut," research and discover the handwriting on the wall and then make their own preparations."

These examples of intellectual orientation in two lives provide insight into how linear reincarnation may conserve views arrived at in one lifetime and use them differently in a later lifetime as the level of knowledge and technology progresses. With such a process, reincarnation would make it possible for the evolution of species consciousness to progress much more rapidly than it would if each human brain started life with no latent memories of previous learning.

Making sense of anecdotal information. Examples like the above that compare a subject's cognitive profile with a previous life can be evaluated with Cerebrotype Rating Scales seen in Appendix 2. The complete Integral Model's case-evaluation package can be downloaded at the Reincarnation Experiment web site.

To rate each life on the five cerebrotype traits, the researcher collects representative samples from biographic data available on both the present and previous incarnations. The best comparisons require material from several sources and from different periods in each lifetime. The information does not have to be as academic as that used above. Both personal comments and behavioral data can be very revealing. Here are two examples.

Laird wrote, "Inside, Marilyn knows deeply of God, me too…" On the overall subject, she writes, "Now I'm just blown away that in our thinking we were the same." Finkelstein found both Monroe and Laird had a "deep belief in psychics and paranormal phenomenon."[7] He also reported that Sherrie was susceptible to self-hypnosis. These data reflect a similar high level of credulity and a supernaturalist mode of thinking in both women.

Where extensive written material is not available, as in the case of Swidersky, we can infer much of a person's cerebrotype if we know enough about his work habits. What we know about Tony makes it possible for us to conclude that both he and Ken are highly rational, disciplined and orderly in their thinking. Tony's nickname "Hackman" came from his proclivity to point out poor workmanship. While not directly responsible, he studied and identified design flaws of the airship he died in. Ken has analyzed flaws in naval and municipal readiness against terrorism.

Having access to and using information like that sampled in this chapter, the experiment has made it possible to compare the cognitive profiles of the fourteen people listed in the Introduction. An illustration of some of the cognitive correspondences can be seen at Figure 17. If the strong scores of cerebrotype similarity recorded for these pairs is duplicated in the other personality-factor and genotype scales, the Integral Model suggests there is a high likelihood that the hypothesized past-life matches may be valid.

COGNITIVE CEREBROTYPE COMPARISONS

Cerebrotype Comparisons. This table illustrates the interaction of the traits and numbers on the personality-factor scales. Each of the five traits represents a continuum between two points, as in the range from a primarily Emotional thinker to a primarily Rational one. A highly rational thinker would be rated five. On the Global/Particular trait, a global thinker would be rated one. The scores are neither positive nor negative; they are simply relative measures of the trait.

The table below illustrates how the subject's scores can be compared to and combined with that of a previous life. The pairs of James-II/James, Lorin /Pitirim, and Ken/Tony have three exact matches each. Those of Kelly/Dolley, Michelle/Mette, and Peter/Paul have two exact matches each. The Charlotte/Lucy pair has one exact match.

Taking into account the subjectivity involved in this scoring system, when the ratings of a subject and a previous life are in adjacent blocks on the same trait they are considered to be a match. The Charlotte/Lucy pair has four adjacent-block matches. Peter/Paul and Kelly/Dolley pairs have three adjacent matches each. The pairs of James-II/James, Ken/Tony, and Michelle/Mette each share two adjacent matches. That of Lorin/Pitirim has one adjacent match.

When both exact and adjacent matches are added for each case, the following pairs have a "high-level" of correspondence on the cerebrotype factor: Charlotte/Lucy, James-II/James, Kelly/Dolley, Ken/Tony, Lorin/Pitirim, Michelle/Mette, and Peter/Paul. Limited data available on Lucy, Mette, and Tony may make their scores less accurate than the others.

Caution: Remember this is just one of five factors that are evaluated before estimating the probability of a past-life identification. Only when this high-level of matches is repeated in most of the other factors can one assume a strong likelihood of a non-random event.

Cerebrotype Traits

	1	2	3	4	5
Emotional/ Rational	LP, PG/PT	C-		DM/K-, KA MG/MM	JM/J2,PS/LK TS
Traditional/ Experimental		TS	KA, LP MM	DM/C-, JM LK MG, PT	J2, K-, PG PS
Reactive/ Disciplined	C-, PG/PT	LP	LK, MG	DM	JM/J2,K-,MM PS, TS/KA
Global/ Particular	PS, JM/J2	K-, LK	DM	LP, PT TS/KA	C-, MG/MM PG
Reflective/ Impulsive	JM, PS/LK TS/KA	DM/K- J2	MG	MM PT	LP/C- PG

Case names are identified above as follows: Charlotte = C-, Dolley Madison = DM, James Madison = JM, James-II = J2, Kelly = K-, Ken Alexander = KA, Lorin Kee = LK, Lucy Payne Todd = LP, Mette Gauguin = MG, Michelle Moshay = MM, Paul Gauguin = PG, Peter Teekamp = PT, Pitirim Sorokin = PS, Tony Swidersky = TS.

Fig. 17

Predictions. With biographical information, the cerebrotype and other factor scales should also may make it possible to predict the kind of person who will likely be the reincarnation of any historical personality. For instance, quotes from the *Declaration of Independence*, largely the product of Jefferson's mind, make it clear that he grasps abstract concepts and deals in general principles derived from a variety of facts that support his conclusions.

"When in the Course of human events, it becomes necessary for one people to dissolve the political bands which have connected them with another. ... a decent respect to the opinions of mankind requires that they should declare the causes which impel them to the separation.

[Jefferson then provides a list of grievances to support it.]

We hold these truths to be self-evident, that all men are created equal, that they are endowed by their Creator with certain unalienable Rights, that among these are Life, Liberty, and the pursuit of Happiness."

Jefferson then describes the expected role of government in securing these rights and makes the case that when a government fails to do so, the people have a right to "alter or abolish it." He asks the signers to "pledge to each other our Lives, our fortunes and our Sacred honor."

When TJ-? is identified, we will likely see similar ways of thinking about the challenges confronting humans as a result of the growth of large authoritarian institutions and corporations unwilling to face the social and environmental costs of their rule. It is likely that Jefferson's egalitarian, individual freedom, and self-responsibility themes will be repeated. As did Jefferson's life, TJ-?'s life will likely include a variety of successful initiatives that demonstrate an imaginative and freethinking mind. He will have probably charted a course of flexible thinking that resulted in breaking with tradition on questions of government, religion, and education.

The model's predictive capability should work in both directions. If we can predict the intellectual modes and interests of the likely reincarnation of Thomas Jefferson, we should be able to equally predict those characteristics for the past life of Brian O'Leary. Brian has the mind of both a scientist and philosopher. His life as an iconoclast has garnered him a combination of praise and scorn for his achievements.

While his science skills have been honed, the philosopher sees and writes about the negative implications of many of society's current technologies and official policies. In the past, he would have been burned at the stake by religious and political orthodoxy.

In fact, that is the kind of previous life that would be expected for Brian. The soul in him today would have animated a comprehensive and independent thinker in the past. During the later middle ages or the Renaissance, that person would likely have been a priest with an interest in natural science and social reform. His writing would have been prodigious and provocative. He would have butted heads with the authorities and placed his well being, and even his life in danger.

Chapter Fifteen

Emotional Egotypes

The Monroe/Laird case introduces this chapter on correspondences found between the emotional patterns of a subject and a hypothesized previous incarnation. According to the case study done by Finkelstein, Sherrie experiences emotional cycles similar to Marilyn's. Each uses drugs and alcohol when depressed. Each is considered temperamental, with emotional flare-ups or loss of control. Even their manner of expressing emotions is eerily evocative of a linear reincarnation. Marilyn, angry and naked at a clinic, threw a chair at someone. Sherrie, in a flash of temper, threw dishes out a window and ended up naked on the lawn.[1]

With low self-esteem, they exhibit submissive tendencies, particularly toward dominant males. However, being ambitious, both express a confident, optimistic attitude toward achievement of ultimate career success. Anxious, both are prone to making snap decisions. Moments of depression are followed by manic initiatives, one cancelling out the other.[2]

As you can see, the objective is not to make one right or wrong, or to define good or bad behavior. In evaluating a past-life connection, it is the degree of similarity in the two profiles that is important. With sufficient quality information about both lives, an independent observer can place each personality on the continuum established for each of five ego traits.

Fortunately, the biographical record on Monroe is monumental and Laird has also created a significant public record of her life. The latter record has been augmented by the research and documentation done by Finkelstein for his book. Sherrie and Marilyn match almost exactly on each trait. Based on the quality of data, the overall egotype match in this case suggests a high level of probability that the psychoplasm incarnated in Marilyn is the same incarnated in Sherrie.

Emotional Egotype Traits

The egotype factor consists of traits that contribute to a human's emotional core. The scale for each trait is like a thermometer; it includes a range of degrees between opposites. For instance, instead of reading from hot to cold, an ego "thermometer" measures a person's position on a behavioral continuum, as from passive to active or from dependent to independent. Where one falls on such emotional gradients defines the energetic stance that I call an egotype factor.

The aim of such personality thermometers is that they help us obtain unique profiles that differentiate among individuals. One can say Jane is different from John in four ways or Lidia is so similar to Ramona in five ways that they appear to be twins. One can then compare the four profiles and predict how an individual is likely to behave even though they live in very different circumstance—or even in different historical times.

With a wide range of apparent reincarnation cases, one should be able to get a sense of the inter-scale reliability. A high reliability score tends to confirm the Integral Model assumption that the psychoplasm transmits the energetic basis for all aspects of the new-born's personality. This means when the current subject and the alleged past-life egotype profiles are very similar, strong correspondences in the other three personality factors are usually present.

The Egotype Factor Rating Scale (see Appendix 3) used in this experiment deals with five traits. The researcher asks the following questions: (1) How cool or hot is the person's temperament under stress? (2) Is the person worried about what may happen next or confident of his or her ability to handle the situation? (3) Does the person always have a high energy level or is he generally depressed? (4) Is the person anxious or calm in ambiguous circumstances? (5) What is the balance between pessimism and optimism?

While one could measure many traits, this combination of qualities provides a measure of how individuals react to their social and physical environment. There are no right-wrong or good-bad answers. The goal is to compare apples to apples in terms of overall emotional traits.

To use the scale, a researcher or self-evaluating individual collects significant biographical data on the subject and the hypothesized previous life. A representative sample of information makes it possible to ascertain where the individual falls on each trait-continuum. With a little bit of practice one can develop a sense of how close the two parties fall on the scale. It is equally important to note significant differences in the two profiles.

Matching Pairs. In another case like Monroe and Laird, with two strong female lives (Gad-Gauguin and Moshay), we see an equally persuasive case for a past-life connection. Mette reacted to her husband's desire to go abroad and paint by optimistically offering her support. She took on the role of promoting the sale of work Paul had done in France and dedicated herself to teaching piano to raise money to support their children.

When it became apparent to her that Peter needed her help, Michelle, like Mette a century before, optimistically shifted priorities to help his art become better known and respected. She also assisted in the research on his reincarnation case. See <http://www.peterteekamp.com>.

Both women energetically and confidently support the priorities of the artists. Each is very stable and engaged once they decide on a course of action. As a woman in the 1800s, Mette had the self-confidence to organize a public art show, particularly one with paintings outside the styles in vogue at the time. With her own career at stake, Michelle demonstrated the same calm nature by facing a skeptical public with her own and Peter's stories of possible reincarnation.

How do the male companions to the two women do on the egotype scale? Both Gauguin and Teekamp express themselves in similar ways. They vacillate between optimism and pessimism. Manic in pursuit of their goals, they become depressed when their lives go badly. Each made at least one suicide attempt when obstacles seemed to be overwhelming.

Paul Gauguin, after running out of funds in Tahiti, returned to France for two unsuccessful years. He returned to Tahiti at age forty-seven. Continuing in poverty and bad health, after two years he tried to commit suicide. Debilitated and poor he died four years later.

During one week's time Peter Teekamp lost his lease on his studio, his girlfriend left him, and an accident totalled his car. He decided to kill himself (at age forty).[3] Fortunately, he took alcohol into the desert with his gun. He passed out from drinking before shooting himself. Learning from it, Peter has now changed the soul's pattern of this trait.

Conversely, confident and adaptable, each of the two men left more financially secure jobs in business to pursue their artistic profession. Cooly putting their own agenda above the needs of family or friends, they prefer detachment to engagement. Both are confident of their talent, driven to succeed in their art.

Each persists against high odds of failure. Gauguin told others, "I shall achieve things of the first order." That did not happen in his lifetime, but posthumously his work became world-famous. Teekamp has the same artistic drive and expects to see wider global acclaim for his already respected works of art.

Describing his soldiers, Gordon wrote, "My men labored as earnestly and bravely to save the town as they did to save the bridge. In the absence of fire engines or other appliances, the only chance to arrest the progress of the flames was to form my men around the burning district... and pass rapidly from hand to hand the pails of water."

Keene, describing the crew working for him, wrote to the department's chief, "Eugene donned full gear and headed toward the rear of the building.... At great risk to himself, he turned off the valves of both tanks.... Christopher displayed good common sense, knowledge and training and deep concern for the safety of the fire-fighters under his command.... It is my feeling that special attention should be paid to the actions of (both)."[4]

How Gordon and Keene reconcile their deep emotional concern for the tragedy of human suffering with the need for confident detachment on duty is illustrated by Gordon's reactions to the battles of the Army of Northern Virginia and Keene's stumbling onto a Gordon battleground.

"As the last words between us were spoken, Rodes fell, mortally wounded, near my horse's feet, and was borne bleeding and almost lifeless to the rear.... General Rodes was not only a comrade whom I greatly admired, but a friend whom I loved. To ride away without even expressing to him my deep grief was sorely trying to my feelings; but I had to go. His fall had left both divisions to my immediate control for the moment, and under the most perplexing and desperate conditions."[5]

When Keene first walked into the battleground of Antietam, near Sharpsburg, Maryland, he wrote, "A wave of grief, sadness and anger washed over me. Without warning, I was suddenly consumed by sensations. Burning tears ran down my cheeks. It became difficult to breath. I gasped for air as I stood transfixed in the old roadbed."[6] In fighting a fire that threatened his men he found the same ability to respond to his professional duty.

These personally written accounts of comparable emotional responses to similar situations suggests the subject and the previous life exhibit the same egotype profile. Both Keene and Gordon were calm and stable in the presence of danger. While cool under fire, they were both passionate about the well-being of their comrades. They remained assertive and confident in dealing with very traumatic circumstances.

The rating scores reveal a high degree of correspondence between their emotional profiles. Combined with similar correspondences on the other factor scales, they point toward a high level of confidence that the Gordon/Keene match reflects the same psychoplasm.

Sorokin and Kee are very close in all the traits. They are both very self-confident, ready to take on the existing political and bureaucratic establishment. Sorokin was anti-czarist in the Russian Empire, anti-Bolshevik after the Revolution. After emigrating to America he was a vocal opponent of the current theories, and personalities, in the field of sociology. Kee has publicly taken on the Federal Reserve and U.S. banking system and is at the center of the push for an independent commission on "9/11."

Both men are on the manic end of the scale, almost driven to get their work done and have it recognized by others, but Lorin is a more "laid-back" character. Each is willing to voice their ideas in as many forums as possible. Sorikin used newspapers, pamphlets, and academic papers. In the twenty-first century, Kee uses the Internet, CDs, and videography as well. Both are split on the optimist/pessimist scale. They are optimistic about their own lives, while pessimistic about the direction of society.

Sorokin's reputation suggests he was a more temperamental personality than Kee, but the latter is also easily riled up about injustices and problems that, as he sees them, ought to be taken care of in a democratic society. While both may be intellectually and emotionally agitated about issues, they calmly and systematically go about their work.

Use of Biographical Material. Since psychological tests did not exist in the eighteenth and nineteenth centuries, we must depend on historians who document terms used in those periods to infer what we now call emotional traits. Melancholy is one such term; sanguine is another. Abraham Lincoln has been described as suffering from deep melancholia. James had an apparent tendency in the same direction. Today we might speak of it as depression.

For this reason we must search through historical accounts for markers of emotional or ego profiles. Exemplary nuggets can be found in biographical accounts and personal papers. With practice, one can identify deep emotional patterns in either the descriptions of the deceased subjects written by contemporaries or in that person's own writings.

For instance, during a review of personal documents for clues to the emotional and interpersonal profiles of Dolley, I found the humor and personal style of her sister Lucy very familiar to me. Reading from Lucy Payne Todd's letters reminded me of communications that Charlotte had shared during the project. Lucy and her older sister Dolley had become members of America's new elite when, as young Philadelphians, they married into leading political families.

At age fifteen in 1793, Lucy married George Steptoe Washington, a nephew of President George Washington. Lucy, George, and their children lived at the Washington family's Harewood House in western Virginia until he died in 1809. (See Figure 18.)

After her husband's death, Lucy lived in the White House with Dolley and James until 1812, when she married U.S. Supreme Court Justice Thomas Todd of Kentucky. Theirs was the first-ever wedding in the new White House.

Lucy and Dolley share the same Quaker background and have mixed feelings about it. Both were disowned by the Quaker meeting (the equivalent of excommunication for Catholics) for marrying out of the faith. Lucy became close to James during her stay at the White House. (James and Dolley had married at Harewood as guests of Lucy

Lucy Payne Washington
(Circa 1798)

Portrait by Charles Peale Polk
(1767–1822)   Fig. 18

and George.) In later letters to Dolley, Lucy fondly referred to James as her "brother." After Lucy's marriage to Thomas Todd, she named their daughter Madisonia and a son James Madison Todd.

Lucy's marriage to Todd and relocation to his home in Kentucky took her away from the excitement of the White House and Washington. In her letters, she pined for the fashionable social affairs and conversations at tables of prominent figures, and her extended family. Years later, widowed a second time, Lucy returned to Washington to live with the then-widowed Dolley until her own death. Her story suggested to me that Lucy might be related to other cases in the project.

Methodology. Each personality-factor scale should be rated separately for the subject and the hypothesized previous incarnation. Replication of the process by an independent observer also gives additional reliability. Ratings on the egotype scale alone do not determine the overall strength of the proposed match. However, when the number of correspondences on one scale is low, it suggests that close attention should be given to the remaining scales.

Only after all the traits in each personality factor have been evaluated and combined on the Ratings Summary Sheet (included in the data pack on the web site) can one make a considered estimate of the probability that the match is valid. A proposed past-life match with low numbers of similarities on more than two scales suggests a weak level of confidence in a valid match.

The Dolley Madison case, with an extradimensional source suggesting her soul had split in this century between Kelly and Charlotte, illustrates how hypothesized matches can yield different ratings for the same past-life match. Based on interpretations of biographies, comments by contemporaries, and personal letters, Dolley can be rated on the egotype scale. With personal writings, psychological self-ratings, and researcher observations over an extended period Kelly and Charlotte can be independently rated on the same scale.

Based on her ordered approach in coping with challenging circumstances, even wartime threats to her personal well-being, Dolley is rated a cool-headed # 1. Given her reactions when challenged to do many things she had never done before, she rates a self-confident #1. While she was often depressed about health, financial, and family problems, she always carried on her expansive life and fulfilled her responsibilities. Her score is #3 on the continuum between the manic and depressive extremes. Facing strife-ridden social situations, balancing that with her personal anxieties, she rates a calm #4. With the many dramatic changes in her life, often soul-rending, she remains a #2 optimist.

Charlotte, from her own perspective, exhibits a different profile. When faced with opposition or frustration she can be a temperamental #5. Worried about her abilities to venture into new territory, she avoids risk taking—a #4. She sees herself as a somewhat insecure, depressed #2, but very changeable person. In positive periods, she produces much excellent work. When faced with uncertainty she becomes tense and impatient, meriting an anxious #1. She often feels pessimistic (a #5) about her personal situation and the general trends in society.

While occasionally temperamental, Kelly usually behaves in a deliberate manner that rates a #2. Given to worry about many small things, she is confident about her ability to handle most challenges and gets a confident #2. She is not easily depressed; even when uneasy about her future performance, she is diligent in her preparations to insure success. This results in middle #3. While worried about details and "possible" disasters, Kelly enforces a state of overt calm (#4) in her work and social life. In spite of her anxieties about details, she remains a #1 optimist about life.

The chart shown at Figure 19 below illustrates how various individuals' egotype scores can be compared in order to quantify the levels of correspondence between lives. For each of the five traits, the initials indicate the egotype ratings for Marilyn, Sherrie, Pitirim, Lorin, Dolley, Charlotte, and Kelly. One can see how the scores for seven people spread across the five scales. They should not be taken as definitive, but the relationships are instructive.

Egotype Factor Rating Scale

Cool/Hot	(1) D	(2) K	(3)	(4) L	(5) M, S, C, P
Confident/Worried	(1) D, P	(2) K, L	(3) M, S	(4) C	(5)
Depressed/Manic	(1)	(2) C	(3) M, S, D	(4) L, K	(5) P
Anxious/Calm	(1) M, S, C	(2)	(3)	(4) D, K	(5) P, L
Optimist/Pessimist	(1) K	(2) D	(3) P, L	(4) M, S	(5) C

C = Charlotte, D = Dolley, K = Kelly, L = Lorin, M = Marilyn, P = Pitirim, S = Sherrie.

Fig. 19

From limited Lucy material, one finds some indicators of her egotype. She is described as "chatty and charming." She is also known to easily sink into "despair." She suffered "deep distress" when her marriage to Todd took her to Kentucky, far from Washington and family. He had to promise to take her back to Washington for at least two months each year. Her vacillating moods result in impetuous actions, exemplified by eloping in her first marriage at age fifteen. Before marrying Todd, she changed her mind several times. In the end, she sent a rider after the rejected suitor already en route to Kentucky to say she had finally decided to marry him.

Charlotte has never married, but she shares some of these traits. Her voluble personality slips easily from excitement to distress. She quickly reacts, but then often reflects and changes her mind. Like Lucy, when in safe situations she sprinkles conversations with her humor and clever witticisms. She is devoted to trusted friends and concerned for their welfare.

This informal comparison identifies more emotional similarities between Charlotte and Lucy than between Charlotte and Dolley. Some of the reasons may be explained as follows.

Reasons for Different Scores. The Integral Model predicts that the psychoplasm's basic traits at the moment of death will carry forward as the platform from which the personality starts development in its next incarnation. Even if true, it is still unlikely that a subject and a possible previous-personality will be rated exactly the same on all factors.

Obviously this process is not perfect. Some patterns in the psychoplasm may not be identified by the five traits on each scale. The evidence chosen may not always be a reliable indicator of the person's real traits. Subjective views cause different scorers to obtain different results. These potential shortcomings must make us cautious about drawing final conclusions.

For instance, with the paucity of biographical data on Swidersky, comparisons of Tony and Ken on the egotype scale must be tentative. However, given their behavior in stressful situations and their approach to challenges in the work place, it is reasonable to conclude they share a high degree of confidence and calm. Their success in the military dealing with disasters would also support a rating of cool. What we know about their similar concerns for details, and particularly those that pose a threat to personal health and safety, suggests both worry about such issues and are often pessimistic about the direction of current developments.

An additional factor that may result in different scores between a subject and his posited previous personality is the fact that phenotypes and personalities continue to evolve at varying rates for different people. For that reason, one must attempt to distinguish between apparent mismatches that result from particular events and variances that reflect a true evolution of the personality. That requires comparisons of how both personalities act in a variety of situations in order to identify the more stable, fundamental trends.

Letters written by Dolley reveal a deferential mode towards men in contrast to a self-confident assertiveness in the feminine role of hostess. She wrote a friend on her wedding day that she expected James to be a "generous & tender protector." Later in life she wrote her son Payne Todd on many occasions in a submissive tone, pleading for his advice on financial matters and his help in solving her problems. Was this behavior a function of the culture or the soul?

Kelly reported an early tendency to seek the patronage of dominant and powerful men. Even while earning a professional salary she still preferred for her husband to handle financial issues. She exhibits a similar deference to her son, being extremely sensitive to his views. Based on this data, both would receive a similar rating on this trait, even though Kelly exhibits many of the independent-ego aspects of the twentieth-century feminist movement.

While personality types do evolve, a warrior is not likely to change into a nursing-home caretaker in one transition. A sycophant in one life will not magically become an independent leader in one leap of the psychoplasm. A person always suffering from insecurity will not likely immediately reappear as a risk-taking entrepreneur. Evolving through the process of reincarnation in a self-learning universe seems to require the discipline of actual learning. Wishful thinking and magical incantations do not substitute for experience-based development.

Chapter Sixteen

Social Personatypes

This personality factor covers the ways in which people interact with others, individually or in groups. It is based on the assumption that how all of us relate to others is more a function of our own inner needs than the kinds of people they are. We shift a bit depending on circumstances, but stay within the parameters of our basic personatype. This interpersonal style evolves gradually and appears to remain stable enough to be recognized over several lifetimes.

Since a deceased person cannot be subjected to modern psychological tests, the researcher uses surrogate information to develop a personality profile. From historical words and actions one can usually infer the traits that influenced behaviors in the previous lifetime. With relevant biographical information from the past and present, one can determine if someone's interpersonal needs and style are consistent with the behaviors associated with a possible previous lifetime.

To illustrate, let's continue with Lucy Payne. After marrying in 1793, she arranged for her mother to live with her and her new husband in western Virginia. Among other options available to each, sixteen-year-old Lucy and her mother decided such a move would be mutually beneficial. In 1794, the widow Payne moved to Harewood, where she lived until her death in 1807. She was present during the early childhood years of Lucy and George's three sons.[1]

After Lucy's husband died in 1809, she lived much of the time in the White House with Dolley and James until she married Thomas Todd in 1812. After the widowed Dolley re-established her household in Washington, the again-widowed Lucy moved from Kentucky to Dolley's home for eight years until her death in 1848.

While economic and social conditions made it common for sisters and mothers to share homes for extended periods in the nineteenth century, Lucy had a variety of options available to her. The fact of her extended adult cohabitation with her mother and older sister is significant. These arrangements clearly met a psychological need which would be prominent in her profile. The fact that Charlotte has resided in the same family home as her mother and older sister for most of her life may suggest a similar trait in her personatype profile.

Continuing with the Lucy story, historical documents describe her as a popular belle at the White House during her extended stays with Dolley and James between 1809 and 1812. Remembering the charms that made Lucy a favorite in Washington society, young Edward Coles once wrote that "nothing but the great disparity of our ages, has prevented me losing my heart."[2]

Lucy clearly enjoyed the social events and travel organized by Dolley and her close friends. Dolley played the role of older sister for Lucy during her stay in the White House. Washington socialites checked with her to make sure their invitations for Lucy were appropriate. "Ruth Barlow, for example, did not dream of asking Dolley's sister Lucy Washington to join her in France without first asking permission of the First Lady." [3] These episodes reveal a younger-sister aspect to Lucy's personality.

Letters Reveal the Persona. Lucy's witty letters illustrate how she saw herself relating to others. She poked fun by embroidering her comments. She used humor in what some would consider a scandalizing way when talking about friends or acquaintances. She also used self-deprecating remarks about herself, revealing that she easily burst into tears. Her jokes had sexual innuendoes and teasing threats of bodily harm or revelations of private matters.

The following examples from her letters to Dolley illustrate this trait. Referring to a friend, she wrote, "... Jack too with his scally nose and wide mouth." Writing of the French Minister she said, "If I disliked him for speaking bad english—he ought to have hated me for speaking no French." On an acquaintance, she described her feelings, "I feel very much provoked at that old hag Mrs. DU.Val.... every body knows her disposition to venom ... tho if I had her head here I'd box her ears." On a friend's love relationship she wrote, "[he should] but take her to france and *finish* her education."

Referring to James's discretion in showing Dolley affection when Lucy lived with them, she wrote, "he was always fearful of making my mouth water—tell him I now get kisses now that wou'd make his mouth run over." On a friend's view of her new husband, she wrote, "Phebes comparing the judge to still champagne was an error in judgment." Writing to Dolley she revealed impressions of herself, "... but I am a bit of a fool, always was and I fear always will be—or rather shall be... for I am almost killed with *ennui*."

Charlotte engages in similar self-deprecating and teasing humor in her communications. Here are some examples that illustrate the trait: About allegations of a companion's misdeeds, she writes, "I would enjoy having a laugh about it together. I was no better myself." After a rough encounter with a friend, she says, "I'm going to throw up now and then go into the bathroom and kill myself." Other times, she exclaims, "UGH! the pictures of me are hideous." "I can't keep my big mouth shut that long." "He has ruined it for me.... moan and groan and moan and groan again."

Though the subjects and circumstances differ, a strong correspondence between the two stands out. They can be evaluated in the context of the Integral Model for interpretation.

Other Illustrative Cases

James has been described as having many acquaintances, but few friends. He is known as "aloof, dour, or disapproving" in public encounters. James-II is known for his "non-emotional, critical, and put-down manner." He, too, selectively develops few close relationships. Seen as stiff and reserved, even cold, neither man has a "hail-fellow-well-met" personality.

Their cool, objective manner, while seen as personally distant, enables James and James-II to be effective in formal settings. James is known to be very persuasive in "argumentation." James-II has a reputation as a capable advocate in government circles. Showing a different private side, both express warmth and joviality among friends.

The former slave Paul Jennings (freed by Dolley in 1840) wrote about James in 1865 when he worked in the U.S. Department of Interior. Born in 1799, Jennings moved with the Madisons to the White House in 1809, where he was a young servant during the War of 1812, and continued to serve James until the morning of James's death on 28 June 1836.

Jennings described James Madison's interpersonal style saying he "never saw him in a passion... never saw him to strike a slave" and that he would not let an overseer do it either. He said when a slave did misbehave, James would admonish him in private rather than have him be "mortified before others." Jennings also reported that if a colored man tipped his hat to James that he would return the courtesy, saying he did not want a Negro to excel him in politeness.

James-II grew up in the mid-century era of racial segregation, but worked closely with poor Negroes in his father's fields. He reports that for no obvious reason he never felt as his peers did and related to the hired hands as he did to his cousins. One U.S. official wrote, "[James-II] works hard at developing and maintaining effective professional and personal relationships with his associates.... He is an opponent of conventional, hierarchical office relationships and is, thus, egalitarian in working with associates. He is, as well, liberal and humane in outlook and, thus totally free of bias." The two personatypes seem to match.

With the extensive historical material available on James and a fairly good set of archives on James-II, it is not difficult to place them on the personatype rating scales. See Appendix 4. Their respective traits fit into the same block or within adjacent blocks. Five such correspondences demonstrate that both likely share the same personatype. If the other personality factors have similar levels of matches, the model suggests a high level of confidence in a past-life connection.

On the five personatype traits continua, James and James-II are rated as follows:

A. Timid/Uninhibited Trait: Both were rated #4.
B. Aggressive/Submissive Trait: Both were rated #2.
C. Dependent/Independent Trait: Both were rated #5.
D. Introverted/Extraverted Trait: Both were rated #3.
E. Trusting/Skeptical Trait: James rated #4 and James-II #5.

With regard to the timidity and aggressiveness traits, as expressed in interpersonal relationships, both James and James-II tends to hesitate and let another person take the initiative in new situations. James took advantage of older people introducing him to new roles.

He looked to his teachers for direction. His father gave him a military commission. George Washington and Thomas Jefferson took him under their political wings. He takes the initiative when empowered by friends and family.

James-II also often waits for intervention by others. He had no long-term patrons, but was befriended by seniors who furthered his career. In meetings he sits on his hands until, like James, he sees a defined role to play. He has bold ideas, but makes few public demonstrations of them. Both men work best in groups. They attempt to guide a group's thinking and bend it to their will, but readily compromise when it would lead to achievement of their larger goals.

In dealing with others, both are skeptical, if not downright distrustful, until they become well acquainted with their interlocutors. Their credulity level has a very high threshold; they want evidence of the other's goodwill and intentions. Both use codes with colleagues to prevent outsiders from having access to their affairs. Overly suspicious, James-II also often projects unwarranted intentions of duplicity on colleagues and adversaries.

In terms of the intro/extraversion trait, each has different styles for public and private circumstances. Both have been described as being more at ease in private circles than in public gatherings. They tend to be solitary in their development of ideas and individualistic in their approach to thinking through problems. Each is introspective and prudent when engaging others in serious discussions.

In terms of independence, neither James nor James-II require much external stimulation to motivate their work. They do not hesitate to express their well-considered views in any forum or media. But, after taking part in a public meeting, each retires to his own chambers to review, revise notes, and decide on next steps.

Evidence for Ratings. One can assess where individuals fall on the continuum of a particular trait by using different sources of information. In rating Dolley and Lucy for comparison with present-day subjects one must depend on the written record of their behaviors. For ratings on people living today we have the advantage of self-assessments and third-party reports, followed up by psychological interviews and access to some personal documents.

Several biographers have indicated Dolley did not hesitate to introduce herself to the world. At age fifteen she arrived in Philadelphia from Virginia and immediately launched herself on the Quaker social scene. She pushed their limits with her dress, social manners, and contacts with boys. After mourning the death of her Quaker husband—who died of yellow fever in October 1793—over the Winter, her good humor and exuberance re-emerged with Spring.

At barely twenty-five, she quickly became one of the most eligible young widows in Philadelphia. In the norms of that era, she was quite independent and uninhibited in society, even introducing new interpersonal and social styles of behavior. She did not hesitate to take the initiative to engage with men in political discussions. Her extroverted manner gained her a wide circle of friends and admirers. She trusted people and they trusted her.

After a somewhat secluded childhood, Kelly became socially active with her college roommates and male friends. Early on she became economically independent, taking the initiative in developing new relationships. In her early twenties she assumed the challenge of a highly visible government position under a powerful mentor. She excelled in the job's personal, office, and public roles. Married three times, and like Dolley, she does not hesitate to challenge the convention of the society's social norms.

Charlotte's career and public behavior remains in the traditional social norms of her community. With a passion for fashionable attire, she chooses to live and dress in a professional manner that does not challenge the status quo. She remains single and lives in the family home with her mother and older sister. Preferring to focus on her private pursuits, she does not exhibit a socially active personality.

When in private groups where Charlotte's status is secure, she is known as a "down-to-earth and funny person." She swiftly responds to situations with humorous and provocative remarks, eliciting laughter. People tell her how much they appreciate her witty contributions.

All three women are trusting, tending to take people at face value instead of being suspicious. Generally forthright with others, the personas people see in public represent the women's actual personalities. But, in private affairs, each keeps her own secrets.

Reliability of the Scale. Adding the Gauguin/Teekamp case demonstrates the reliability of the personatype rating scale. First, with two similar people the scale should result in matching measurements, as shown above in the James/James-II case. Second, the scale should differentiate among individuals for whom different psychoplasms have been hypothesized

That means the measurements for Gauguin and Teekamp should reveal a similar profile, but it should significantly differ from the James/James-II profile. Third, the Mette/Moshay ratings should be different from all of the four men. All three of these expectations are met.

Gauguin and Teekamp are venturesome travellers. Gauguin moved from France to Denmark, worked in Panama, spent time in Martinique and the Marguesas Islands, and made Tahiti his artistic home. He also travelled the world as a sailor. From a childhood in the Netherlands, Teekamp made his home in the United States. He also travelled to India, Israel, Egypt and Portugal and toured the world painting murals.

Both live very individualistic lives, spending much time in solitary pursuits. Even when not travelling solo across the globe, each is very much a loner. Each is suspicious of authorities, spouses, colleagues, and ideas contrary to their beliefs. Teekamp, orphaned in childhood, became independent in his teens, plying his art in the hippie/communal lifestyle of the 1960s. He could not get a situation much closer to the nineteenth-century artistic colonies that helped stimulate Gauguin's work.

Both Gad-Gauguin and Moshay are venturesome and bold. Mette moved to France from her native Denmark, became fluent in French and managed a household. She smoked cigars and dressed outside the female norm of the late nineteenth century. Michelle wanted to travel alone to Tahiti and France as a young woman, but acquiesced to her parents wish for a female companion. She is now an individualistic spokesperson on the value of objective evidence for reincarnation.

Both women tend to be impulsive, animated about their current project (supporting their adopted artists or pursuing their own personal goals). Michelle and Mette exhibit sympathetic and considerate tendencies in their relationships to Teekamp and Gauguin respectively. Michelle combines her extroversion with success in marketing and promotion of her ideas in public media. However, she appears intimidated by authority figures. Mette is not evaluated on this factor.

The four individual souls involved in these two apparent linear reincarnations are joined over space and time by common traits. Each believes in a principle beyond themselves: the power of art to capture and convey soul-filled images to others and to transcend generations. Michelle and Peter, not married to one another, are dedicated to being examples of the continuity of consciousness in art and personalities in their own life of self-development. They do this, as Mette and Paul did a century before, by daring to live outside a conventional "box."

Other Examples. The following paragraphs consists of the kind of data from which one can infer the personatype traits. Sorokin and Kee similarly question established authority, hold skeptical attitudes about conventional beliefs, and are determined to live their own sense of self. Sorokin's early years were characterized by subversive political activism first against the Czars, then the Bolsheviks. For this, he was jailed often and nearly executed. His criticism of the machinations of government were published in books and his newspaper *Regeneration*. Even after Bolshevists raided his offices and destroyed all of his presses, he kept on writing and publishing, playing cat-and-mouse with the authorities all the while.

In the current era, Kee feels we are living under a different, but still tyrannical government today. In recent years, his "revolutionary activities" have increased significantly. As an avowed anarchist and member of the "9/11" truth movement, he has spread DVD copies of *Loose Change* and *America: Freedom to Fascism* all over America. He puts them in paper sleeves and leaves them on community and college campus billboards and in airports across the country. His web site is a clearing house for this kind of "citizens, open-your-eyes" information.[4]

During this experiment, he sometimes questioned why, if things went so badly for him during a previous lifetime, he has the tendency to fall back into the same activities again. He has concluded that if he ignores the spirit within him, it would be like turning his back on what he believes is an important part of his soul purpose.

On the other hand, Kee feels that another part of the soul's purpose is to find ways to improve on several attributes in the Sorokin incarnation. Sorokin's work at Harvard was brilliant, but he was not one of its best-loved teachers. He was viewed by many students as conceited and arrogant. With colleagues he dogmatically held to his own theories, often belittling the works of others. Kee takes this to mean that he needs to practice more humility this time around, critiquing his own ideas as well as those of others.

Both Swidersky and Alexander sought the structure of military careers. Though both demonstrated a need for this hierarchical system, both are skeptical enough to question their superiors. Both tend to submit themselves to routine and established roles; Ken followed a routine of regular fasting and prayer through his twenties. Both are more introverted than not, with Ken remaining celibate until his marriage and both of them marrying in their late twenties.

Sherrie's e-mail communications provide excellent insights into her personatype. She wrote her reincarnation therapist Finkelstein the following: "Hi...! How are you? Well I thought I would try to get a lot done today. I really have been in the habit of shutting off or out the whole world and staying awake very late and sleeping in. ... my biological clock is such that I am with great ease a night person and with great difficulty a day person. ... Were they interested in our story? Were they believing in this kind of thing?"

She continued, "Anyway I wrote so many great e-mails to films, funds, documentarians, and magazines. Nothing back yet. I'm a very impatient person when I write out something. ... I will try for my Canadian citizenship. Hah - unless we can find someone in the U.S. to marry me. Hee! Hee! I don't know. Do they still do stuff like that? Imagine the most famous American female can't get back in!" These spontaneous sentences express an uninhibited boldness, with self-reliance and impulsiveness (like Marilyn).[5]

Aid in Personal Development. The primary purpose of the factor-rating scale is validation of specific past-life identifications. However, it could also be a useful aid for self-reflection. Regardless of whether one is able to validate a specific past-life identity, use of the scale to rate oneself in the present can provide insights into what is likely to have been brought forward. Identifying unexpressed needs or conflicts with today's circumstances may reveal tensions between a soul legacy and aspirations for growth in this lifetime.

For instance, in the review of similarities between Dolley and Kelly, an apparent trait appeared to persist and call for wider expression. Dolley gained a reputation as a grand hostess in both the Jefferson and Madison White Houses, but she never had the opportunity to establish herself fully as the "mistress of her own house." She reportedly got along well with her mother-in-law, but Nelly Madison was the grande dame of Montpellier until six years before James's death. By then his health and their financial difficulties had derailed Dolley's dream.

With no conscious awareness of this possible past, Kelly has focused, from her first marriage, on establishing a home that allows free range to her strong desire and considerable skills for high quality entertaining. Her family's situation and her early schooling did not include models or training for such a role; she intuitively expressed an innate predisposition bolstered by inherited skills. Three husbands, including James-II, recognized her deep-seated need and tried to honor this motivation and obvious capability as James had done more than 150 years earlier .

Self-guided, past-life research using the Integral Model approach may provide a better understanding of our own internal conflicts. Use of the factor scales to clarify whether one's life decisions are internally consistent or at odds with some inherent aspirations may help identify alternative, and perhaps more soul-based, behavioral choices.

A third-party researcher, if in the role of therapist, coach or teacher, can use the same insights to provide helpful feedback for a client. The scientific researcher can use partial comparisons to help the subject explore other areas of evidence to see if a stronger case evolves, or if the broader search points toward a different past-life.

Chapter Seventeen

Creative Performatypes

Hiding with her family from the Nazis in Amsterdam, Anne Frank wrote, "I know I can write." But she asked herself, "Will I ever be able to write something great?" In December of 1943, while confined in the annex behind her father's office, she decided to give something funny to each member of her family. She attempted creative writing by "composing a little poem for each person." With its Holocaust roots, her diary posthumously became one of the most widely read in the world.[1]

Anne was fourteen when she wrote her "little" poems. By the time Barbro Karlen was twelve she had written her first book of poems that became the best-selling poetry book ever in Sweden. In five years she had published nine more volumes of poetry and prose. In this pairing of creative performatypes we see talents from a previous life manifested more fully expressed in a new life. Did the soul of Anne experience in the young Barbro the thrill of being a wildly successful teen-age writer?[2]

All the stories of prodigies and precocious children discussed in Chapter One may be like the Anne/Barbro story in one significant way. Each of them apparently has his or her roots in one or more preceding lives. What if we researched the lives of long-deceased people who had talents similar to our current prodigies? Would we find similarly corresponding performatypes, and all the other psychoplasm factors, between some of them and Terrence Tao, Jay Greenberg, Daniel Tammet, Dalton Roberts, Willie Nelson, and others?

How do we identify performatype prints in a particular lifetime? A person's professional label or job title may not be the same from life to life or one's career category may be vague enough to cover a variety of creative activities. To compare apples to apples, as done in the other three personality factors, I use a simple topology that divides into six categories the occupations that are found in all human cultures. See

Appendix 5. In this context the word occupation includes creative work wherever it is done, paid for or not.

Soulprints in Children. Although kindergartens do not administer vocational interest tests, the reincarnation hypothesis predicts we could identify vocational predispositions in early childhood. Psychologists have not developed such assessment tools because they assume children don't develop such preferences until much later. However, pre-school children, like young birds ready to quit the nest, apparently carry imprints that shape early behaviors and choices that end up being vocational paths.

One Mettler family story has young friends making fun of George as they played around his house. Before he and his friends were old enough for school, George spent much time sitting on the front porch reading newspapers. He had no idea where his habit might have come from until, during our research, he looked at the portraits of Johnson with him always reading newspapers. Did George start this life with habits well honed in the past?

Several of the Stevenson and Tucker cases mentioned elsewhere indicate vocational patterns like George's were more than precocious; they were of pre-birth origins. Imad loved to repeat his previous life's hunting activities. Sukla and Jasbir started this lifetime dedicated to their Brahman religious practices. Swanlata was a dancer from the start. Parmod could have become a baker had he wanted to follow his past.

To test this assumption of early interests one needs only to observe the play and self-chosen past-times of the very young. Unless coerced by overly eager parents, a child's choices differentiate it from the predispositions of other family members. Though exposed to similar environments and opportunities, each child chooses his own path.

For an infant to know what it apparently knows about its interest and abilities would take years of learning if it had to start without a pre-existing base. Noam Chomsky (discussed in Chapter Two) unwittingly pointed to such a pre-existing base when he theorized a child's early use of language without specific instruction depends on innate linguistic patterns. The cases in this book strongly suggest that more than linguistic principles are inherent to the new-born.

Clearly Teekamp was born with artistic skills. He reported to me that when he entered school, without training, he wowed his classmates with pencil portraits they treasured. His precocity also matches Gauguin's mature style. After Michelle had begun to take seriously the possibility of her own past-life connection, she found several Gauguin pencil sketches that resembled some of Peter's youthful drawings.

Several years later, while Semkiw was researching the Gauguin/Teekamp case, he asked Peter to paint a few pieces in Gauguin's style. Although he had not consciously liked or imitated Gauguin's style, Peter agreed and surprised himself more than anyone by the flow of art that could be Gauguin painting new versions of his own work for different clients.[3] Peter, unaware for most of his life that his work paralleled in any way that of Gauguin, has now demonstrated the carry-forward of an inherited performatype legacy.

Gad-Gauguin and Moshay share the general artistic factor with Gauguin and Teekamp. Gad is a musician (piano player and teacher) and Moshay is a poet, actor, pianist and writer. However, their personal performatypes also differ from those of the two men. But the ways they differ make the two women similar to one another.

Moshay's conventional trait appears dominant in the vocational area. She has worked as an administrator and business person in the field of newspaper advertising. Gad shows the same trait in her efforts to market Gauguin's art and selling her services as piano teacher and childcare provider. Given their nurturing of the two volatile artists, both women also share the social traits of generosity and service to others.

In areas less obvious than painting or music, it is more difficult to identify similar occupations in two different eras. For this reason, instead of asking what kind of jobs a person has or desires, the researcher's questions elicit information about underlying needs or preferences. This psychological information then points to the kinds of jobs or hobbies that will be fulfilling.

The Sorokin/Kee case illustrates different manifestations of an investigative or research trait. When Sorokin was an impoverished young man unable to pay for university classes, he supported himself as an artisan and a clerk to study at the Russian University of St. Petersburg. His psychological motivation for achieving advanced knowledge was strong enough to give him the energy and focus to create a way to pursue his studies. Kee, similarly motivated, with a young family to support precluding graduate school, secured a job that gives him access to classes.

In his video production work he spends long hours taping lectures in philosophy, political science, sociology, psychology, and history in some of the finest universities in the country. The representative of one of his client schools jokingly—but more significantly than he knew—declared that Lorin should receive honorary degrees for all the classroom time he logged.

On the performatype scale, Sorokin and Kee share three categories. Their most important one is the "investigative," based on their curiosity and the motivation to learn described above. Each, Sorokin vocationally and Kee avocationally, scientifically pursue their intellectual interests through their research and analysis. They then use various media to make their findings public.

The second shared factor is "conventional," because of their attention to detail and valuing of precision. While Sorokin worked in a university, due to his unconventional theories, he was hounded to the periphery of his own profession. Kee decided to remain outside academe while he also challenges it, but does his work on his own turf as an independent scholar.

Both can be placed in the "artistic" category for their love of music and writing. Sorokin's affinity for music is clear and his writing is legendary. Kee tried his hand at piano and school bands. His creative writing was considered among the "best of his college."

The connection between internal, psychological traits and the creative work we choose makes it possible to compare past-life linkages in spite of different vocational circumstances. The existence of the same traits in different specific jobs has been illustrated by Sorokin and Kee. Let's now look at correspondences between James and James-II that suggest the same performatype even though their titles and the scope of their responsibilities were quite different.

The interests and values that led James to be successful in institutional building during the early decades of America's new government appear to surface again in James-II's activities as a career government official. In facilitating the Constitutional Convention and later in the Congress and Administration, James worked on and wrote about the constitutional and organizational issues involved in creating a democratic republic.

In a later century, James-II observed from a distance and wrote his evaluation of President Nixon's efforts to correct some of the problems that had developed in the structures and processes James had helped put in place. James-II, as an official participating in a similar reform process initiated by President Carter, attempted to make sure the strategies to reshape federal organizations made sense in historical terms. He wrote recommendations, private and public, tried to shape the Office of Management and Budget reorganization plans, and urged Congress to revise what he considered unworkable institutional changes. He found himself writing—in books and articles in the modern era—on some of the same issues that James had faced.

Obviously James and James-II share with numbers of their respective contemporaries a desire to create a political system that provides for as much self-government as possible. People who share that goal include warriors, clerks, farmers, businessmen, nurses, politicians, and most other occupational types. What makes these two men different from the others is the cluster of factors that led them to assume the roles they did in their respective political circumstances.

Both are curious about the social and psychological dynamics (James didn't use that phrase) that shape communities and societies into nations. They read as widely as possible the relevant history and commentary on these issues. They analize the information they gather and build theories that apply to the situation facing them. With attention to detail, they carefully write and communicate their findings and recommendations. Valuing organizational effectiveness, they devote their well-developed minds to creating appropriate institutional structures.

In another example of different-jobs/similar-performatypes, the careers of Ken and Tony reflect the same career symmetry. Both married in their late twenties and grew up in landlocked towns based on energy industries, but joined the navy. Serving on shore at naval air stations, neither went to sea as they worked in technologies new to the Navy. Tony worked on the new technology of lighter-than-air flying ships and Ken was one of the Navy's first environmental engineer officers. Both also had technical civilian careers: Ken in civil engineering and Tony in the rubber-making industry.

In performatype terms, Ken exhibits the same realistic traits as Tony. Both joined the Navy even though neither wanted to be a traditional seaman. They liked the opportunities it provided to deal with things material and mechanical. Both have conventional values of attention to detail and orderly processes, combined with an investigative curiosity and bent toward problem solving. Chemistry and engineering are common interests. On the rating scale they had three common factors, which is a level of correspondence suggesting a likely match in this factor.

Ken's life may illustrate a soul's journey that develops new levels of expertise and skill from aborted experiences in a previous life. A very bright enlisted Tony died before reaching his potential; Ken received graduate degrees and achieved the rank of Navy Captain. Tony died in the largest Navy aviation disaster of all time; Ken was able to develop the Navy's first formal disaster management program after the 11 September 2001, terror attacks in the U.S.

In the alleged Denver/Kern shared-soul case, John and Steve share three factors: the artistic, the enterprising, and the social. This level of correspondence could be interpreted as support for the possibility of a shared soul.

In addition to being a musician, John soared as a singer and composer. He was also an actor, writer and poet to fill out the artistic factor scale. When in college he studied engineering and architecture, suggesting an interest in the realistic factor scale. He was very enterprising in the development of new projects and programs to promote his social, political, and ecological values.

On the social factor scale, John undertook humanitarian and charitable activities and worked for peace with Russia and China during the height of Cold War era. He co-founded the Windstar Foundation to promote ecological values and the development of citizens to take responsibility for building a peaceful and sustainable future.

Steve demonstrates a similar range of interests and their expression in lifelong activities. In his music and work in the arts he clearly manifests the artistic factor. The enterprising factor stands out in his career of starting projects, movements, and companies. He has also run conventions and produced events of many types.

Steve's social factor shines in his public education and training programs, work in nonprofits, and support for environmental causes. He is of the peaceful Buddhist persuasion (as was John) and works for international peace in many ways.

Power of the Past. The psychoplasm hypothesis assumes that creative patterns are developed and elaborated over many lifetimes and are robust enough to imprint themselves in the new incarnation. This means we should be able to predict from previous patterns what the likely directions of creative expression will be. While there may be shifts in relative priorities, it is highly unlikely that a new life will leave behind all of the top two or three strongest areas and commence a radically new direction.

Preliminary evidence from this experiment indicates the new incarnation takes one or the other of two options. In option one, the individual continues to develop the specific priority of the previous life. In option two, the person shifts priorities within the same cluster of interests.

Physician Norm Shealy continued and built upon the physician cum metaphysician legacy carried forward in the psychoplasm from the life of English doctor John Elliotson (1791–1868). Elliotson introduced the stethoscope and the use of narcotics, as well as mesmerism which he learned in France. He also called into question the hypocrisy of medical science. Shealy achieved renown for his development of new pain-control technologies. He has also been interested in the realm of parapsychology and intuitive healing. He, too, published a novel on the hypocrisy of medicine. In these ways his career has furthered the Elliotson path.

The James/James-II case illustrates a reordering of priorities in middle age. He followed the early educational, military and political inclinations seen in James's life. For most of his life, James was always able to serve in self-directing roles as legislator, political appointee, and elected executive. As a career official, James-II first managed to have positions that gave him leeway to pursue his innate values and priorities. When that became impossible, he founded his own private company to escape the bureaucratic culture. This step allowed him to pursue the intellectual interests James could not fully develop, even after his retirement.

When the official part of their lives ended, James and James-II pursued private studies, created educational opportunities for others, and served society as a volunteer. James served in the Agricultural Society and on the founding board of the University of Virginia. James-II helped found and lead several nonprofit cultural and professional societies. James worked to consolidate a written legacy of his ideas and career. James-II is developing a synthesis of his experiences.

The Monroe/Laird case illustrates a path that keeps the artistic focus but allows the current incarnation to try out variations within the same area. Although Marilyn Monroe is best known for her acting in many movies, she wanted to be recognized as a singer. While Laird has the same general artistic performatype as Monroe, she has focused on developing her skills as a singer and musician. With less attention to acting and modelling than that given by Marilyn, Laird seems to have stressed the musical facet of the psychoplasm's artistic repertoire.

The Gordon/Keene pair illustrates another case of slight adjustments made by a person to remain congruent with the soul's priorities. Following the early vocational pattern of nineteenth-century soldier John Gordon, Jeff Keene served a twentieth-century stint in the U.S. military. When Jeff's first tour of duty was over, he moved into a career as fireman to satisfy the soul's penchant for a military-like organizational structure and social environment.

Ingrained performatypes seem to carry a degree of force to surmount local barriers to their self-actualization. Keene had to overcome institutional opposition in the fire-fighter profession to have a career consistent with Gordon's success in the military. Gordon rose from civilian soldier to one of the most noteworthy generals in the Confederate Army. With dedication, Keene followed suit in a comparable career in a fire department's structure of ranks and hierarchy to become very successful.

Both men, after establishing their bona fides as an officer and leader of men, became active in more individualistic endeavors that challenged them to develop other aspects of their soul's performatype. Gordon became involved in educating the public on the value of a new political order. (His path took him to the Reconstruction-era governorship of Georgia.) Keene took on the task of educating the public about the value of a new level of consciousness about reincarnation.

Sorokin and Kee perceive their destiny to be the development of their intellectual capacities in the study of human institutions and cultures. As a child Sorokin began a regime of self-teaching—supporting himself as artisan and clerk to gain admittance to the best Russian universities—that eventually led him to a professorship at Harvard in 1930. Kee, in another time and set of circumstances, also pursues his intellectual development through self-teaching while supporting his family working as a video editor and DVD developer. A similar determination in pursuit of higher goals motivates both men.

Vocation Versus Avocation. Another way a soul can develop different strands of an overall packet of creative interests is through avocations. If financial conditions require that one's vocation be limited to a narrow area, the strong inherited interests will be expressed in one's avocations, as in a hobby, volunteer work, or recreation.

Charlotte was born with artistic talent, expected by some to become a commercial artist. She imagined planning special social occasions with tasteful and joyful decorations. She naturally organizes meetings and projects at work and for groups in which she is involved.

Dolley was a successful organizer of social events with political impact. With tact and warmth she made people of all political persuasions comfortable, setting the standard for gracious American hospitality. Serving as unpaid national hostess for the widower President Jefferson and her husband James, she established the role known later as the "First Lady."

Kelly and James-II also privately hosted Capitol Hill dinners and salons reminiscent of popular Madison gatherings in Washington and Montpellier. However, Kelly was able to turn Dolley's avocation into a paying vocation. On a less visible but analogous scale, she was hired to use her similar talents and interests to organize international political and social gathering for a nonprofit in Washington, D.C. Perhaps drawing on a Dolley psychoplasm's organizational skills, Kelly easily succeeded as an administrator in the higher education and nonprofit communities.

Like John Adams, James apparently considered a career in the ministry while in college. His tutelage under Presbyterian minister John Witherspoon (president of Princeton), including an extra year after graduation reading Hebrew among other studies, would have made it easy to enter the clergy. James-II was formally ordained as a youthful minister. On an informal basis, he served as pastor of a small church and later in a spiritual manner consistent with his evolving beliefs.

Following additional avocational tracks, after college both James and James-II considered, and privately studied, the profession of law before they ended up in the practical aspects of government. Concern for the role and meaning of religion in society continued to be an avocation throughout their adult lives. One of the foremost champions of religious freedom in America, James promoted the separation of religion from civil government in the Virginia legislature and in the drafting and ratification of the U.S. Constitution.

Much of James-II's avocational writing has dealt with the same issue of separating personal religious beliefs from the policies of government at all levels. His work in the area of religion appears to have been an (unconscious) attempt to take James's Enlightenment perspective to a new level of a natural spirituality. Thus, we see the same performatype at work in avocations consistent with the context of two different eras.

The Garden Path. Dolley and Kelly have a lifelong avocational engagement with gardens. When they had the space, and help, each designed and nurtured mixed flowers and vegetable gardens. Dolley created a beautiful rainbow design in her garden at Montpellier, with slaves to keep it fresh. Charlotte says she dreamed of having a beautiful garden, but with someone else to do the yard work. This characterizes all three who seem motivated by the aesthetics of a lovely garden more than by the grub work of digging, planting, weeding and keeping the predators away.

The examples in this chapter illustrate the case for a strong performatype carry-over from one life to the next. The many correspondences of creative interests reappearing in two or more lifetimes add support to the hypothesis of sequential incarnations by a single soul. If these beyond-chance levels of correspondence are backed up by high quality and valid information about all the five personality factors, it is reasonable to assume a high level of confidence in a specific past-life identification.

Life-path Applications. The performatype scale is designed to identify correspondences between a person's present vocational and avocational activities and those of a possible previous lifetime. However, it may be useful to individuals as a framework to self-evaluate the congruence between their current activities and the needs of their core personality or soul, if you prefer.

The performatype may be used to make a private projection of the kinds of work or play that would fit more comfortably with one's inner sense of self. Used in conjunction with the other factor rating scales, such a performatype evaluation could help individuals make career changes that would be more appropriate for their long-term development.

Predictions. The performatype factor should make it possible to predict the character of a hypothesized reincarnation of a known historical figure. Despite a reputation as a political revolutionary, Thomas Jefferson was also a shy, soft-spoken man with a love for music, architecture, natural philosophy, and education. More of these traits in the soul's essential self came together in the founding of the University of Virginia than in any of his political activities. Will the TJ-? sought during this experiment reflect the same performatype factors as Jefferson?

If he takes the first option described above, he will be directly involved in a political career. If he chooses number two, he will focus on one or more of the other four creative areas. If he is truly like Jefferson, TJ-? will likely feel pulled in several directions simultaneously. We should look for a multifaceted individual who mixes all of them in a new and unique pattern.

Chapter Eighteen

Coincidence through Memory

Most of us think of a coincidence as an accident, an event that happens by random chance. However, sometimes we feel an uncanny sense that it may have an invisible connection to something else. Referring to this phenomenon, Heraclitus wrote in 500 BCE, "The unapparent connection is more powerful than the apparent one." If we see no obvious explanation, we conjure up one: God's will, luck, angelic intervention, or the new non-locality principle now espoused by many physicists. Which explanation we choose depends on our worldview.

The results of what Heraclitus called "unapparent connections," Carl Jung called "synchronicities." He used the term to suggest that such events have some meaning beyond simple chance. He did not mean that one event necessarily causes the other, only that there is some unknown connecting link. Some interpret such events as the work of spiritual beings.

For instance, writing of his own hypothesized past-life link to John Adams, Walter Semkiw called attention to the fact that he had worked for an oil company named Unocal 76, with "The Spirit of 1776" as its slogan. He suggested this was evidence for his possible previous lifetime as President John Adams—who was obviously involved in the American "Revolution of 1776." He also put in this category the fact that he signed the contract for his book *Return of the Revolutionaries* on John Adams' birthday.[1]

Referring to these events and similar examples in other cases, he wrote, "I believe these things represent the working of the spiritual world, of spiritual guidance and assistance on our earthly plane." Reiterating that theme in a later book, he added, "synchronistic events are not accidental... but predetermined." He then concludes, "Through symbolic coincidences, the spiritual world sends messages to us."[2]

From this perspective, coincidence is seen as a supernaturally manipulated process that serves advanced beings (ABs) instead of a natural connection. Accepting a supernatural-being explanation requires we must first establish the authenticity of the ABs and, secondly, prove they caused the events. Read the following paragraph and imagine a nonhuman being looking down and deciding to pull all the necessary strings in Keene's life to make him act as he did.

On his eighteenth birthday in 1965, Jeff Keene reported to an induction center where he volunteered for military service. While the officials took everyone else who applied that day, they sent him home to return another day. Thirty years later, researching his own reincarnation case, he realized that he had volunteered for the military and been sent home on his birthday 100 years after the cessation of the war that had consumed Gordon and the nation.[3]

From a natural science perspective, I assume organic laws at work in the universe do not include divine beings manipulating such mundane movements and decisions. Principles governing human consciousness can account for these events. They do not require a third-party to insure that Jeff, on his birthday, gets to the recruitment office and then to manipulate the minds of the officials to reject him—to make a point that has no impact on him until three decades later.

The postulate that such connections are made by a soul-based memory that encompasses at least two lifetimes has a basis in everyday life. All of us use symbols, objects, special dates, or mnemonic devices to help recall images and facts related to this lifetime's experiences that are important to us. Would a two-life memory not do the same?

The matching of birthdays, ages, and locations with certain types of events do seem to play a subconscious role in decisions that mimic earlier lives. Chapter Six noted that the letters Dolley and Kelly respectively wrote to James and James-II that led to their marriages originated in Hanover County. Add to that the coincidence that James and James-II were both about forty-three years old when they received the letters. Could a two-life consciousness be at work?

Jeff wrote that on his thirtieth birthday, 9 September 1977, he had to go to an emergency room due to sharp pain from his right jaw through the neck and out at his shoulder. Years later he learned that a thirty-year-old Gordon had been wounded in the same spots during the September 1862 battle on the Sharpsburg Sunken Road.

James-II bought his first home on Constitution Avenue, years before he considered any past-life connection to the Father of the Constitution. James's first private home in Washington, D.C. was on "F" Street and almost two centuries later James-II unwittingly bought a second home on "F" Street. Kee named his son Alexander before he was aware that the "A" in Pitirim A. Sorokin stood for Alexandrovich—which indicates Pitirim's father's name was Alexander.

Such coincidences may have meaning. However, I prefer to assume that nothing more complex than a human's natural use of such touchstones to maintain a sense of self needs to be invoked. It is not supernatural that the soul reminds itself today of things it did in another life.

This "soul-memory" explanation is consistent with Rupert Sheldrake's concept of the extended nature of the human mind and Dean Radin's notion of entangled minds (further discussed below).[4] These concepts suggest that Jeff's consciousness interacted with the draft board officials' minds on his eighteenth birthday. Perhaps their sensing of his internal emotional and soul-based state about going into another war caused then to reject him.

Chapter Nine describes the case for a psychoplasm that coheres, maintains, and carries forward more than just the patterns contained in the ten percent of our DNA known as the genome. (The genome DNA is only encoded for and used by the body to make its vital proteins.) I argue that the psychoplasm—perhaps using some of the other ninety percent of the body's DNA—is encoded for and predisposes our mindsets, the stance of our egos, our interpersonal style, and our creative expressions. Embedded in this theoretical package are the codes for knowledge and expertise developed over many lifetimes. The codes include memories, tastes, and habits that remain active and useful to a new physical incarnation.

If we are connected to past-life memories in a manner that is not unlike the way an adult is connected to childhood memories, it is understandable that Walter, Jeff, Kelly, James-II and Lorin would, in their deeper levels of conscious awareness, pay attention to slogans, birth dates, anniversaries, and addresses. That memories from previous lifetimes might result in the above coincidences should not surprise us.

If one psychoplasm animates a previous life and a present life and makes some choices between lives, we should be able to discern the effects of similar choices in each life. Could one choice, that of a new mother, made by the psychoplasm that animates Charlotte have been made partially on the basis of Hare being her new mother's maiden name? Could the Hare maternal name link her to Harewood House, Lucy's first married home where her mother lived with her?

Could a similar desire for symmetry in two lifetimes influence choices about birth order? Dolley is the first-born daughter of her family, while Kelly is the older of two children. Dolley's sister Lucy (the Payne family's second daughter) was born about nine years after her. Charlotte, the second daughter in her family, was born about nine years after Kelly. Do these involve chance similarities, or could they be psychoplasm choices? If so, do they lead to personality effects?

Finkelstein documents many events in Sherrie's life that suggest memories of Marilyn's life may be activating them in this life. Marilyn had a crush on Tony Curtis. As a young adult, Sherrie dated a man who looked like Tony Curtis. Both Marilyn and Sherrie, at an early age, married the army guy next door. Sherrie felt drawn to travel to Banff, Alberta before she learned Marilyn had also visited there. Many more such coincidences abound in this and other cases.

Unapparent connections in science. That events like those above do not have obvious physical connections did not escape the attention of scientists in the last century. Physicists in recent years have repeatedly proven that all physical entities are not as separate as they seem. At the subatomic level, the appearance of solid and separate particles disappears. Instead of discrete boundaries divided by space, patterns of connectedness show up everywhere.

Albert Einstein's view of the universe provided for these connections, but offered no testable explanation for them. Reflecting on this conundrum, he called it "spooky action at a distance." When a pair of photons is split and separated by great distance, action on one simultaneously causes the other to react. That such relationships inexplicably extend throughout space and time revealed that one of their features is "non-locality."

Describing the "near" aspect of these "at-a-distance" relationships, Erwin Schrödinger coined the term "entanglement." As one of the founders of quantum theory, he assumed the quantum mechanics operating in our singular and integral universe were characterized by this image. Dean Radin elaborated on the implications of this now obvious, if not fully understood, aspect of physical theory in his book *Entangled Minds*. He synthesized this and other concepts into a coherent explanation for many psi phenomena. I would add reincarnation to the list.

Another relevant phrase from the world of mathematics and physics is "strange attractors." It refers to numbers, particles, and line trajectories that appear to skip around randomly, but from another perspective (as in fractals) seem to demonstrate order in apparent chaos. In reincarnation research these strange pairings seem just too coincidental to be considered the results of random chance. Seen in two lifetimes, the connections beg for an explanation.

How do we get from the physics of the microcosm to human-to-human or soul-to-soul interactions? A "physics of consciousness" may be the answer.[5] The conscious field of a psychoplasm that manifests in a human would be subject to all the above scientific concepts. A coherent field of memories could not remain stable over time in a single life without the force of strange attractors to hold them together. For these fields to interact over distance would require non-locality. The transfer from life to life described in Chapter Nine would require entanglement. The psychoplasm described in this book is consistent with these new concepts in physics.

The Twins Analogy. Identical twin studies may help illustrate how the physics of consciousness with its "coincidence-through-soul-memory" may work. I use a set of biological twins only as an analogy to a subject/past-life pair. The twins are recognized as two people who share a common parental genome. The reincarnation pair is assumed to share a soul genome.

Most readers have likely heard about research with adult identical twins who were separated in infancy and reared in different circumstances. Scientists have found that beyond the expected common physical features, twins share similar personality traits, mannerisms, job choices, attitudes and interests.[6]

These twins frequently report they share common tastes in clothes, foods, pets, and home decorations. Their spouses or children may have the same names. Often they may have similar favorites in music, hobbies, and even vocations. Some find they visited the same vacation spots or collected friends much like their sibling's acquaintances.

When one examines the number of such twin coincidences, it becomes clear that it exceeds what would occur just by chance. No one suggests the twin linkage is imagined or orchestrated by an extradimensional being. We assume, and scientists statistically argue, that the twin-genome overlap (even though we don't know exactly how it works) results in overlapping fields of consciousness in a way that appears the two minds operate along the same or parallel tracks.

This chapter illustrates that even more correspondences than those between twins can be found in two unrelated personalities separated by death. These past-life correspondences should attract as much scientific attention as the examples published in twin studies.

Coincidences in Reincarnation. The anecdotal similarities often found in reincarnation cases fall into the same areas discovered in the identical-twin studies described above. In addition to predictable physical and personality traits like those covered in previous chapters, the reincarnation subjects also have similar physical habits, tastes, attractions, and aspirations.

As you read the examples in this chapter remember they are sample illustrations from the experiment's files. All strong reincarnation cases are filled with such coincidences. Alone, they do not prove a past-life match, but the sheer volume of them makes the linear reincarnation case more persuasive. As in twin studies, the parallels do not stop when the child grows up.

Habits or Tastes. As habits (patterns of behavior) developed in childhood carry into our lives as adults, it appears many habits developed in one life appear to extend into another. There are at least two categories of such patterns. One consists of body postures, use of hands, speaking, and other physical actions that we unthinkingly repeat. The other involves behavorial patterns that we engage in without conscious awareness. They include tastes for food and drink, music, sex, and various rituals.

Conducting this research made me aware of one of my current habits ingrained earlier in life. Now in my sixties, when dressing each morning I give a military tuck to my shirt, whether it is formal or not, and align its buttoned edge with the edges of my belt buckle and fly. When the alignment is completed it perfectly meets the navy boot-camp standards I learned in my early twenties. Some of the following examples appear to be habits learned in a previous lifetime.

Semkiw has pointed out in his description of the visual evidence for the soul of Wayne Peterson having also been incarnated as the Venetian Francesco Foscari (1373–1457) that they shared posture and the way each posed their hands.[7] Jeff Keene was surprised when he saw in several photos of John Gordon that they both had a tendency to always stand with their arms crossed in front.[8] He also discovered an even more unusual habit they shared. Instead of putting folded papers or other items into the shirt pockets, both tucked them inside their shirts.

During the experiment, James-II and Kelly laughed when they read of reports by the Montpellier slaves that James would frequently slip up behind Dolley to "frighten" her in a chair. The result would be a round of chasing one another across the veranda. They said it resembled the teasing antics they use as expressions of affection when trying to get the other's attention.

Personal Tastes. Manner of dressing is an important way in which we differentiate ourselves from others. Since it is often quite idiosyncratic, similarities between the two lifetimes may strengthen the case for a specific match. As an example that involves two couples from different eras, both James and James-II are habitually conservative in dress. In contrast, Dolley and Kelly always outshine their drab looking husbands. They literally blossom with fashionable dresses and accessories in public. They seem to others to be larger than their husbands.

Is this solid proof of one soul being incarnated in two lifetimes? Obviously it is not, but further details do make this case more intriguing. Paul Jennings, James's manservant, said James was "neat, but never extravagant with clothes" and that he always wore a black coat, breeches, and silk stockings, with buckles for his shoes and breeches. He said James kept only one dress suit at the time, because he wanted to be an example of conservative probity to poor relatives.

In his own career James-II opted for conservative dark blue and grey suits, one of each. He believed that one should "wear out one set of formal clothes before buying another one." He has always needed only a small closet for his entire wardrobe. Anecdotes like these from two lifetimes may reflect a psychoplasm's tastes that endure over time.

Reinforcing the intimation of a link between the two lives is another illustration of their desire to set examples of probity: Jennings reported James liked to cut his own firewood at school for exercise and whenever he was home on the plantation. James-II was taught to cut firewood by his father and continues to supply his own fireplace with wood he chopped when possible. James-II often confounded his staff by doing many menial tasks along with them.

In their years as very public younger women, both Dolley and Kelly had a penchant for provocative forms of dress. They wanted to be stylish and did not hesitate to be flamboyant. With a "handsome bosom" Dolley loved to wear high-waisted, low-cut gowns. She crowned her dresses with exotic turbans of bright colors and plumage. As a young wife, Kelly also enjoyed low-cut dresses that spotlighted her figure. In Kelly's professional appearance in Washington almost two centuries later, she followed Dolley's interest in couture with well-designed suits and dresses accented by brightly colored scarves, all supported by the best quality shoes.

One aspect of Dolley's dress habits was the wearing of colorful turbans. When Dolley's hair began to turn white, she tucked bits of dark hair under the edges of the turbans. Equally concerned about appearance, Kelly, in her turn, takes advantage of modern hair coloring techniques. To cover grey roots, she has them dyed monthly to maintain her ageless head of hair. Dolley wore scarves to cover neck wrinkles, while Kelly prefers turtleneck blouses.

Music and Rituals. One of Sorokin's greatest loves is the choral music of the Russian Orthodox Church. Kee has always loved sacred choral music and considers Rachmoninov's *Vespers* to be some of the most beautiful music ever composed. Sorokin wrote extensively on the role of religion in society, judging the world's major religions to be "supremely edifying." Despite this, he rarely participated in religious rituals himself. After shedding the beliefs instilled by his Christian upbringing, Kee, too, turned his back on the rituals of organized religion.

The University of Virginia cases involving children include many examples where the subject was drawn to and able to sing or dance to the music of the culture associated with the putative previous life. If we paid more attention to all young children we might be able to make inferences about their soul tastes. This argues for early exposure to a variety of cultural stimuli to provide possible comfort for children whose previous incarnation was in a different culture.

Eating and Drinking Habits. In addition to the carry-over of preferences in clothing, books, art, housing, and furnishing, it appears one's attitude toward food may survive as well. As soon as she could talk, Barbro told her mother about her distaste for brown beans. She said "I really ate myself sick on them the last time." While in hiding for over two years Anne Frank had only brown beans to eat for many meals.

During his years of service as the new America's Minister to France, Thomas Jefferson became quite enamoured of the French cuisine, which was so different from the bland tastes of his English tradition. He collected recipes and learned to prepare them. Back in Washington and Virginia, his official and private tables influenced his friends, including James and Dolley.

The Madisons' table at Montpellier was as cosmopolitan as finances would allow. As might be expected from the earlier evidence, James-II and Kelly choose to have international menus when possible. Once again, the model predicts for a TJ-? a taste for the exotic and eclectic cuisines of foreign countries.

Paul Jennings reported James was "temperate in his habits," adding, "I don't think he drank a quart of brandy in his whole life. He ate light breakfasts and no suppers, but a hearty dinner [called lunch today] with one glass of wine. When others drank, he touched a glass to his lips or diluted his with water." In the soul's current life, James-II's consumption of alcohol is also light, even in heavy-drinking military and political circles. Neither has been known to smoke.

Relating to their eating habits, when entertaining at formal meals, both Dolley and Kelly preferred to be at a central point of the table. Regardless of protocol, their husbands learned the value of ceding it to them because it complemented their habits and contributed to their own political success. Both women make all feel comfortable in spite of political or social differences.

Entangled People. Examples of coincidence described up to this point have involved memory carry-overs and the non-locality principle. But, the phenomenon of entangled minds also seems evident in another type of coincidences: Cases where unrelated people seem to have tuned into the memory field and personality traits of the subjects.

For instance, the first position in Washington to which James was appointed by a President (Jefferson) was to serve as his Secretary of State. As James-II's somewhat parallel track was developing, the first position in Washington to which he was appointed by a President (Johnson) was in the Department of State. This decision, made possible by several people in Johnson's administration, eventually led to the involvement of James-II in other activities that paralleled James career. If chance was not the cause, it required the entanglement of minds.

When James, age thirty-six, returned to the Continental Congress, he called for a revision of the Articles of Confederation. About the same age, James-II testified before the twentieth-century Congress on the funding of U.S. government programs and how to improve effectiveness. Based on his experience in Nixon's and Carter's government reform initiatives, at about age forty James-II wrote a major report on why they were not successful. By the age of forty, James was involved with the Bill of Rights, the constitutionality of a national bank, and Congressional compromises over funding for the siting of the new capital of the United States.

Such mind entanglement may draw people together in a way that reinforces awareness of soul connections. Over a twenty-four-month period after Steve Kern accepted a possible link to John Denver, he "accidentally" met a dozen people who had had varying ties with John during his life. Included were musicians who played with John, professionals who had worked with him in other areas, people who had provided various kinds of support, or were casual contacts.

While some of them were the kinds of people one might expect Steve to meet in his own musical circles, the others were not. Adding these experiences to similar connections by members of Steve's family (discussed in the next chapter), and given the five-factor correspondences described earlier, these coincidences require an explanation more persuasive than "by chance."

I believe such things can be reasonably explained by the physics of consciousness, but why do they occur? Do such apparent connections show that our subconscious minds playfully leave soulprints as benchmarks to be recognized among our soul cohorts? Do we use them to remind ourselves of the consequences of our own past lives? Or, do they simply reflect repetition of established patterns? Some form of intention is likely involved.

Intended Coincidences. Much has been made recently in consciousness research about the power of conscious intention. Its impact can be clearly seen in the deaths of four of our Founding Fathers. John Adams and Thomas Jefferson, active in the Revolution and later the second and third U.S. Presidents respectively, both died on 4 July 1826. As key players who helped launch a new vision of democracy in the New World, departing this life together on the fiftieth anniversary of the Declaration of Independence was seen as a clear signal of soul intent.

This hypothesis gained credibility when James Monroe also died on the 4th of July in 1831. The fifth President, he had helped Jefferson foment the independence movement in the Commonwealth of Virginia and fought in the war with George Washington. As James Madison, the fourth President and the "Father of the Constitution," approached death some urged him to hang on until 4 July 1836 (60 years after the Declaration). But he decided to die peacefully in his chair on 28 June instead, telling his niece it was nothing more than a "change of mind."

If we were able to identify the current reincarnations of all of these four souls, if they exist, would we find similar examples of the power of their intentions? Would it be possible to identify such influences on some of their decisions in this lifetime?

A short time prior to James's death in June 1836, he was visited on his death bed in Montpellier by Charles Ingersoll who was the first to call him "the Father of the Constitution." Searching for a place to retire, James-II and Kelly chose a home in a locale that lies about the same distance from the Blue Ridge Mountains as does Montpellier (although in a different state).

After moving in, James-II read the town's fathers had designed its central square with a courthouse to be completed in the summer of 1836. The square's inaugural plaque reads, "The Public Square embodies the rights guaranteed to the people in the Constitution of the United States, including Freedom of Speech and Peaceable Assembly." An act of subconscious intent?

Unrealized Dreams. When someone develops an interest in a particular topic but is unable to satisfy its desire for knowledge and experience during that lifetime, could it be possible that the soul seeks to complete the task in the next? Here again, the James/James-II pair offers a hint that the lack of fulfilment motivates action in the next life.

In 1803, James, as Secretary of State to Thomas Jefferson, directed an initiative to acquire New Orleans and Florida from Napoleon. The result was more than expected: The purchase of New Orleans and the French Louisiana Territory. The planned Lewis and Clark expedition to the northwest took on new meaning. Jefferson and Madison's vision, shared by Dolley, for the expedition's journey of exploration and discovery had more than political objectives.

Both Jefferson and Madison, amateur scientists all their lives, had a keen interest in the region's Indians and their languages and customs, its minerals and topography, and its flora and fauna. As members of the American Philosophical Society, the oldest learned society in the U.S., Jefferson, with James's avid support, requested Lewis and Clark to collect botanical and zoological specimens to lay to rest wild speculations about this unknown wilderness. They were also likely motivated by a desire to counter the eighteenth-century assertions by French naturalist Comte de Buffon that all Western Hemisphere species were inferior to those of Europe.

For Jefferson and Madison, Lewis and Clark served as vicarious researchers in a journey they could not make in person. The explorers named the Jefferson and Madison (and Gallitin) Rivers that form the Missouri headwaters in Montana in honor of their benefactors. Specimen collections returned by the expedition must have filled James with a scholar's pleasure, but his thirst for knowledge about the West must still have been unfulfilled at the time of his death.

Terminating the Washington, D.C. segment of their journey in this lifetime, James-II and Kelly moved west, near one branch of the historic Oregon Trail. With a sense of some destiny and avidly reading, they drove across country following traces of the mid-nineteenth century pioneers from Independence, Missouri, to the Pacific Northwest.

Along the way, they discussed the role Lewis and Clark had in introducing this huge section of the continent to the early American mind. They noted the rigorous and often fatal travails of families who had made the trek across the Louisiana Territory and beyond. While many have retraced the Oregon and Lewis and Clark routes, did the journey's unexpected psychological impact on these two people have any special significance?

They spent much of the next five years climbing the mountains and trekking the deserts of the Northwest. They rafted the wild rivers. They took their guns into the wild, avoiding harm to wildlife, to play at being rugged pioneers. They camped and studied the animal life in various ecosystems. They explored the seacoasts. They befriended descendants of both European settlers and Early Americans, seeing how the two cultures had merged. Could this have been the Madisons' dream come true?

Final Caveat. Lists of synchronicities or coincidences like those described in this book's sample cases cannot by themselves be considered definitive proof of a past-life connection. They are not unimpeachable evidence of a psychoplasm transfer from one life to another. People not interested in reincarnation—but reincarnated nevertheless, if the theory is valid—would also have such coincidences and not recognize them. Therefore, the coincidences themselves could not be considered as acts to make them aware of a previous life.

The best explanation for the number and variety of coincidences identified during the experiment appears to be the carry-over of memories in conjunction with the same personality traits in two lifetimes. An integral reincarnation model would predict that such preferences, predilections, or interests will manifest in choices and behavioral patterns in successive lives linked by a transcendent psychoplasm.

Chapter Nineteen

Cohorts in Consciousness

The central focus of the experiment reported here has been an evaluation of evidence for linear reincarnation in general and the identification of specific past-life connections. However, it appears that souls do not reincarnate in isolation. Several researchers have noted cases where groups of individuals who are connected to one another in this lifetime also appear to have had connections with some of the group in a previous life.

Stevenson and Tucker have studied subjects who had twin, other-sibling, or close-family relationships in their alleged former lives. While all relationships do not repeat, many people seem to return in the same family or as members of the same village or tribe. In these cases of family or community ties, the emotional connections remain tangible and vibrant.

Semkiw in *Return of the Revolutionaries* identified a number of possible reincarnations of people who worked together during the American Revolution. Some of them believe they have returned together for a specific social purpose. Others believe they have been able to become more conscious of those previous lifetimes in order to help document the reality of reincarnation for the public. Semkiw also tentatively identified soul groups from other historical eras.

This "cohort hypothesis" postulates that souls who live in the same era and interact in joint endeavors (social, professional, political or otherwise) may decide to return to human form about the same time to continue to evolve together. They may hope to re-establish connections similar to those experienced in the earlier lifetime or simply to participate in the same zeitgeist.

The Teekamp/Moshay case exemplifies two people whose previous lives involved marriage to one another but now may have relationships to further work unfinished in the past. Michelle is instrumental in publicizing Peter's art (as Mette did in the nineteenth century) and uncovering his connections to Gauguin. She discovered that Peter's early sketches (done before he knew of Gauguin's work) resembled very much some of Gauguin's art. Her personal networking led to the discovery of a lost original Gauguin that Peter recognized and authenticated.

Several self-identified cases, such as Jeff Keene, have discovered friends or colleagues who appear to have been in the subject's previous life. Jeff believes three of his fellow fire-fighters may have also been comrades-in-arms during the Civil War. While their cases have not been corroborated to the extent that Jeff's has, the physical correspondences suggest more detailed evaluations should be done.[1] Identifying such a group as a soul cohort would be consistent with Semkiw's hypothesis.

The argument for the reincarnation of soul cohorts makes theoretical sense. At the core of my natural model is an ongoing process of soul experimentation and learning. That suggests each lifetime ends with a residue of "unfinished business." Among many things still in process could be unresolved issues among people just left behind. For instance, intellectual and ego struggles among the people engaged in the American Revolution were not all resolved upon their various deaths and some may wish to work on them in this lifetime.

In order for the souls to learn the lessons of that era, it would make sense for them to continue to play out their conflicts until they learn to resolve them effectively. Conversely, it would be just as logical for souls to celebrate successful achievement of self-assigned developmental objectives. In either case, if souls have some control over when and where they reincarnate, it would be reasonable to choose a setting where they might work on them together.

A Fuzzy Picture. Both tentative theories and evidence in this area are less self-evident than in the individual linear reincarnation cases described thus far. The lure of a self-fulfilling prophecy becomes powerful; a person whose individual case has not been well-established feels more confident if others claim to have shared that previous life. If one believes himself to have been a Founding Father, to know someone who also claims to have been his eighteenth-century colleague can reinforce that belief. But two such mutually reinforcing beliefs do not equal positive proof.

The evidence reviewed here suggests varying degrees of intimacy may persist over lifetimes, with relationships taking on different forms. In addition to cohorts, such connections have been variously called affinity groups, cultural or issue groups, dipoles (two souls with specifically opposing agendas), soul families, soul siblings, soul twins, and even soul-splits.

What these terms mean depends on the researcher and, often, on the nonhuman voice that makes the claim through a human channel. Keep in mind the problem of translation of such extradimensional information discussed on page 116. Subsequent human conversations about these hazy ideas result in further confusion, depending on the assumptions of those involved. To support the soul-cohort hypothesis, we must independently verify such fuzzy clues.

To make cohort groups credible, confirmation of their relationships should be based on an evaluation of the same categories of evidence involved in individual past-life identifications. Phenotype correspondences, cognitive patterns, egotype evaluations, interpersonal traits, and modes of creativity should be developed for all members of the supposed group. Each profile should be compared to the assumed past-life identity to map out the overlapping relationships.

Soul Mates. The term soul-cohort grew out of the idea of soulmates. That concept arose in Western culture from a story by Aristophanes in Plato's *Symposium*. It suggested that the original humans were divided by Zeus into halves and that each person seeks its other half. This eventually led to the romantic belief that everyone has a perfect mate somewhere.

Between the non-intimate soul cohort and intimate soulmate ideas lies the concept of soul families. Being a member of a "soul family" implies a group of mutually attracted souls who have been biologically or legally related at one time. They may or may not share a developmental interest and may have no common purpose beyond a shared family environment. They may also differ from cohorts who see themselves together to share significant evolutionary experiences.

The label "soul siblings" implies two or more souls were birthed by one set of parents in a previous lifetime. They could have been ordinary siblings, fraternal twins of one or both sexes, or identical twins. In this lifetime they are not biologically related, unless they are incarnated in different roles in the same family tree. Cases of siblings in all these categories have been identified by the University of Virginia's Division of Personality Studies.

Stevenson described the case of one woman who gave birth to fraternal twin girls who were believed to be the same souls as her parents. One carried memories and physical features of her grandmother and the other carried memories and features of her grandfather. In another case, two identical twins were born after the death of older siblings in the same family. Though twins in this life, their habits linked one with the older deceased girl and the other with the younger.[2]

Evidence reviewed in this experiment does not support the idea that two people who have been biological fraternal or identical twins in any lifetime have shared the same psychoplasm. The data actually suggests the contrary: While identical twins share the same set of chromosomes, they do not share a psychoplasm. As with fraternal twins, each identical twin has its own unique development path.

Two different twin cases from the university files described by Tucker illustrate the primacy of the psychoplasm over the parental genome. In the first case, a set of identical twins—Indika and Kakshappa—reported two very different previous lives. The alleged previous life of Kakshappa was that of an insurgent against the government of Sri Lanka. If valid, this case demonstrates a separate, inherited psychoplasm for each identically biological twin.

In the second case, two identical English twins also demonstrated the principle of separate psychoplasms. Tucker reports persuasive data that suggest they were the respective reincarnations of two older sisters who were killed in an accident before they were conceived by the same parents. Each had knowledge associated with a different sister, and one had the physical marking of the deceased sister whose memories she seemed to have.[3] It seems to be established that being twins does not mean having the same soul.

In the absence of evidence for shared psychoplasms, it would be inappropriate to create a category known as soul-twins. That concept could also be confused with what some people refer to as a soul-split, where two people who have had no apparent biological relationship are alleged to have been created from different parts of one soul. One is alleged to have inherited part of the past-life soul genome and the other the balance.

Soul-Splits. This concept has been used by psychics, channels, and regression therapists who have received information from extradimensional sources that leads them to believe that two or more humans can now share a soul who in the past lived as a single individual. To explain this concept, some draw on the analogy of biologically identical twins who share the same set of chromosomes. In this case a soul, like a zygote, would split into two genetically identical halves.

This would mean that one disembodied psychoplasm with its energetic container of psycho-physical patterns would have to replicate itself prior to two upcoming acts of physical conception.[4] One version would entangle itself with one zygote and the other with a zygote in a different woman's womb, perhaps years later. A process of replication (splitting) and waiting, if it does occur, would happen in a dimension that we cannot validate through scientific means.

I do not try to prove or disprove the soul-split concept in this effort to assess the scientific evidence for the integral-reincarnation model. Readers may review the extradimensional information supporting that concept in several books listed in the Bibliography. In this section, I address the soul-split concept in the context of apparent evidence for soul families.

On the basis of the above argument that undermines the idea of soul-twins (biological or otherwise), the biological-twin analogy for soul-splits appears to be faulty. If the concept were correct, we would be able to identify the two splits through their shared patterns in the factors that link them to the same previous personality. I have seen no evidence of two humans living contemporaneously who share the same profile as their reputed, previously shared soul.[5]

In a universe where so much is still unknown, we can assert any notion we would like. However, I believe it prudent to ground new theories in the reality we can test. In this case that means not positing an exception to the principle that governs the self-replication of all other organisms. Self-replicating cells maintain and transfer all of their genome by first replicating its DNA, and insuring its integrity, before cleanly splitting itself into two new cells.

The research described in Chapters Eight and Nine has expanded traditional concepts of consciousness to include non-locality, mutual entanglement, and multiple dimensions. This leads some to think each individual mind/soul/consciousness can be anywhere, doing whatever it would like, in however many life forms it desires. This state of metaphysical ambiguity and the lack of a common vocabulary may lead to confusion.

For instance, that human souls A, B, and C obviously reflect some similar traits may make it difficult for nonhumans or psychics who perceive on the energy level to appropriately label the relationships among underlying psychoplasms. When nonhumans beings explain such subtle degrees of spiritual connections as entanglement and non-locality, their English translations into terms like soul-splits, core identities, and personality aspects can only result in confusion.

This state of affairs suggests the term soul-split may reflect a misunderstanding of cases that naturally fit into the relationship types found in a soul family. Soul-split could have been created to explain an overlap between the lives of two people who are actually sibling souls, but perceived on the basis of limited evidence to have been a single soul in the past. Thus, the label may have been precipitously introduced in specific cases before the evidence was examined.

Personal Ambiguities. One effect of the confusion surrounding the concept of soul-splits is the psychological confusion it creates in the person concerned. To label someone a soul-split makes it difficult for that person to develop a clear vision of what his relationship is to that previous personality. With no knowledge of how the division was made and which parts went to each split, the individuals are frustrated trying to identify which of their present traits came from the alleged previous-personality. They have no clues to the origins of their traits that are not attributed to the alleged soul that split itself to make them a part of who they are.

Charlotte, who is one of the core cases in the experiment, finds herself in just such a situation and agreed to have her view made known in the book. As described in the introduction, she was first identified as the reincarnation of Dolley and then, a year later, as a soul-split of Dolley. During this experiment, personality evidence surfaced that could link her to the life of Dolley's sister Lucy. Faced with three possible identities, she had to sort out her feelings.

She wrote a regression therapist to learn if she had had any experience with the soul-split phenomenon. The therapist responded, "... that parallel lives, where our soul is 'split' into holographic portions of energy, have been reported by a number of... clients during hypnotic regression." And then said, "This is also true for numerous regression therapists..."

Charlotte says her "only 'certitude' is that something we call reincarnation is taking place for all of us." When asked if she favored the split concept, she replied, "I do lean a bit in that direction, but the jury is out as far as I'm concerned." She emphasizes that she has only reported what the AB sources have indicated and that she never says, "I was Dolley Madison." She says her current personal perspective is that "I may have been Dolley, Lucy, or neither."

A Shared Soul? Another proposed soul-split case is based on parallels in the lives of the deceased John Denver and his still living contemporary Steve Kern. John was born with an Air Force father on 31 December 1943 and died in a plane crash on 12 October 1997. Steve was born with an Army father on 11 November 1957. Steve is also a composer, musician, and singer, but did not particularly follow John's musical style until after John's death.

Now Steve is often asked to do John Denver tributes, sometimes as "Little John." (See the web site at <http://www.tributesabroad.co.uk/tune/info_denver_littlejohn.html>.) One spectator's response was something like, "Wow! It's sort of eerie looking at Little John. He looks and sounds so much like John Denver, you think John Denver is performing right in front of you."

A YouTube videographer decided to pay homage to rock stars who had died tragic deaths. He mistakenly selected a photo of Steve Kern to put on his John Denver page of honor. When asked why he used that picture of Steve he replied, "It was the best one I could find of John." The two pictures in Figure 7 show why a fan could have made the mistake he did. They are "look-alikes," but do they share the same soul from the past?

A comparative study of the lives of John and Steve with the Integral-Model factor scales reveal a number of similarities. Steve, like John, has always been an avid environmentalist and international activist for world peace. Their singing voices are somewhat similar.

John, a rebel from parents and institutions, had a difficult youth. Described by a biographer as a loner, he left home for California, to return later to Texas to complete high school. He dropped out of college after taking engineering classes and courses in music. He moved again to California to pursue his musical interests. There he had two police arrests, worked with two different groups, and then struck out on his own. He married, divorced, and married again. The success of his music, with 300 recorded songs—one-half he composed himself—is history.

Steve also took music courses in college, but switched from a music major to social work. His work career has been as peripatetic as John's in its early years. Married and divorced, his personal life has been as turbulent as John's. Each fathered one child. After using his business and entrepreneurial talents in a number of projects, Steve now focuses more on his role as a musician.

The six facial geometry measures described in Figure 8b reveal definite differences in bone structure and facial geometry. The body types tend in different directions—John's toward a taller mesomorph and Steve's toward a shorter ectomorph. While phenotype differences do not point to a single soul, they may be compatible with a birth-family connection in a previous lifetime.

On that basis, I decided to look at the personality factors. The correspondence scores are close enough that they may be hints of a close soul-cohort relationship. The two men could be members of a birth-soul family, perhaps siblings who shared a general orientation and similar interests in one or two lifetimes.

Testing whether that might be the case, I looked at other members of both families for overlapping circles of interests and relationships. I found other Kern family members involved with John Denver's family and/or the Windstar Foundation co-founded by John. The following actions by members of Steve's family could be consistent with an extensive soul-family network. That Steve did not facilitate these connections suggests that it is actually an inter-family affair.

Steve's sister-in-law is president of the Northern California Windstar Connection (where Steve's brother and niece are also active). She is a friend of John's brother Ron who is now president of the Foundation. Steve's sister traveled to Ireland with John's mother and a small group. Steve's niece is named Jenny and a nephew is named Zachary, the same names given John's children. His brother's second wife's surname is Delaney, the same as John's second wife. Given the model and evidence covered in this book, the two families appear to be a soul cohort.

Soul Cohort Validation. A well-documented study of a possible cohort—like those proposed by Semkiw—would provide a credible public test of the psychoplasm and the general concept of reincarnation. Reports of the initiative could attract self-developed cases that would add significantly to the data base of verifiable evidence. With enough cases, the soul-cohort concept and individual past-life identifications would be seen as worthy of scientific study.

A robust set of individual cases in the context of a self-identified cohort could lead to public support for an adequately-funded research program to develop a general reincarnation hypothesis. Well-documented cases for a group of living individuals who clearly manifest the traits and behaviors as did their reputed past-life counterparts should interest the public.

Let's assume the study shows that the individual cases in the alleged cohort *and* the group's current dynamics are proven to reflect the historical record. The weight of those combined findings could make it the most persuasive case in history for a theory of reincarnation.

Demonstrating the impact of inherited legacies on the way people actually live today might help people understood that even if history does not repeat itself, the same personalities do. The lesson might be that if we do not make progress in our personal and social development in this generation, the next generation will repeat the same mistakes that led to today's problems.

Future Explorations

Throughout the book, I have focused on testing the psychoplasm concept through tangible evidence. With its posited integral and linear transfer of psycho-energetic patterns from one life to another, the psychoplasm accounts for all areas of evidence generally related to past-life matches. I believe it is a plausible explanation for the most common images of the soul.

Even if this basic, soul-genome concept, with its genotype and personality carry-forwards, stands the test of time, many questions will remain, requiring credible answers before we can develop a widely-supported general reincarnation hypothesis. People who seek answers to these questions often turn to metaphysical techniques and sources that cannot be verified by objective means. These efforts are natural evolutionary steps in gaining new human knowledge.

Some of the most interesting, and scientifically intriguing, questions are already being addressed by a variety of metaphysical groups. The Bibliography contains several sources that can bring the uninitiated reader up to date. My hope is that for the benefit of our society, scientists and metaphysicians will soon become willing to sit down in open-minded discussions of questions like those below. A cooperative effort could open new areas of human knowledge.

Various notions of karma have often been associated with beliefs in reincarnation. Science can help demonstrate how natural-cause-and-effect principles in everyday life may also play a role in what some call karmic consequences. A joint effort could start with the question, "How does the Sanskrit notion of karma as the 'result of action' play out over multiple lifetimes?"

An issue that might be resolved from joint research by scientific and metaphysical groups is, "How do we determine what, if any, credence should be given to nonhuman sources and the various methods used to communicate with them?" Answering that question would facilitate efforts to determine the reliability of such sources and to validate their information.

"Does the individual psychoplasm have any choice over the time and location of its next incarnation?" Evidence discussed on the Reincarnation Experiment web site suggests some degree of freedom at the individual level does exist. "How much control does the psychoplasm have over its transition from the present life?" A central question is "What happens between lives?"

In order to understand the possible impact of previous lives on today's attitudes and actions, we need to address several questions: "What determines, if any, the relative shifts in personality traits from one lifetime to another?" "Does it matter when and what a person learns about a possible previous life?" "How much can an individual change the legacy of a past-life?"

Credible answers to these and other questions about the millenia-old concept of reincarnation could cause a shift in both the consciousness and behaviors of people around the world. However, before most people will begin to explore and take seriously this uncharted area of the human experience, they must see the evidence of its effect in daily life.

The well-known and reliable trance-channel Kevin Ryerson referred to earlier has an expansive vision of the implications of widespread acceptance by Western culture of the role of reincarnation in life. He believes it could be useful in medicine, psychology, and career counselling. He notes the idea of recurring themes in sequential lives would provide predictive possibilities for choices in many areas, including one's participation in political activities.

Ryerson already believes that reincarnation is a proven phenomenon and that we can use this natural human capacity for the betterment of individuals and society. He invites people to look at societies and cultures who already make open use of reincarnation ideas. He notes that on a daily basis, many cultures practice cultural rituals that demonstrate its validity in the same way that our scientific experiments keep our [Western] worldview intact.

He asks, "What would the world be like if doctors, engineers, artists, and architects worked with such a worldview?" He hopes the reincarnation concept will infuse daily thinking, rituals, and technology. He believes "we are approaching a time when our understanding of complexity will result in a natural synergy supportive of... the re-imaging of our society and culture."[6]

While I would not use the word "proven" as Kevin does, I now believe that a mechanism somewhat like the psychoplasm evolves—as does an individual biological cell—and sequentially shapes the members of each new generation of humans. I feel confident that we are on the right track to understand this phenomenon because the information in this book about individuals, past and present, can be verified. Biographical information from their lives can be corroborated by third parties. The interpretations of this data can be re-examined by other researchers.

Though opinions about what it means may differ widely, the corpus of evidence stands on its own. For those whose worldview depends on the logic in nature, this compilation of specific correspondences—links between the past and the present—points to a "genome of the soul."

I believe the model of an information-rich psychoplasm provides a testable concept within the current state of our understanding of the natural universe. Until a better explanation for the evidence renders it obsolete, further experiments with this model can be a productive step towards a general reincarnation hypothesis.

While I cannot say how, I believe the evidence suggests that—in some form or the other— the learning from previous lives provides each of us with an innate legacy to build upon. It does not matter who or what we might have been in a previous life. We cannot change that. What really matters is to consciously develop in this one the legacy each of us would like for his or her soul genome—for lack of a better term—to energetically transfer into a new-human form.

Acknowledgments

No one writing today on the subject of reincarnation with a scientific orientation can begin without honoring the research legacy of Ian Stevenson (1918–2007). From this point forward, every scholar must take into account his unimpeachable and monumental body of evidence. While he might not agree with my extrapolations, the Integral Model would not have evolved without his foundational data. My evaluation of fundamental traits as predictors of cognitive, emotional, and behavioral correspondences is grounded in the seminal work of David McClelland (1917–98). I hope to add to their pioneering work in the deepening of human self-understanding.

This book would not have evolved as it did without the early support and encouragement of Bob Friedman, president of Hampton Roads Publishing. I am grateful for Hampton Roads' publication of my earlier books and their two books on reincarnation referred to throughout this text. Walter Semkiw, author of one of those books, added to Stevenson's multi-faceted data base in ways that stimulated my developmental-psychology perspective and pointed to aspects of this psycho-physical, five-factor model. I appreciate his mold-breaking work.

During the development of the data and the writing of the final manuscript, I benefitted from the substantive comments and editorial suggestions of a number of colleagues. While I took all their views very seriously, in my reactions, I may have veered from some of their intentions. For that reason, I am solely responsible for the outcome. Although not a complete list, I would like to thank particularly (in alphabetical order) Tony Adams, Susan Fincher, Tom Hansen, Tonette Long, Jeff Mishlove, Norm Shealy, Howard Pepper, Jim Tucker, and Wilja Witcombe.

A large number of participants in the pilot project and the experiment described in this book were very generous in sharing their personal histories and views on reincarnation. Some deserve special recognition for entertaining and responding to my many inquiries into matters one normally considers to be very personal. The openness of Ken Alexander, Lorin Kee, Jeff Keene, Steve Kern, Sherrie Laird (through her revelations to Adrian Finkelstein), Michelle Moshay, Peter Teekamp, and the individuals in the anonymous cases made possible a validation of the personality-based approach to identification of past-life connections.

The book would not have materialized without the graphic design expertise and dedicated efforts of Lorin Kee and the skill and commitment Susan Wenger applied to its composition and layout. Wheatmark Publishing merits kudos for its commitment to a tight publication schedule to meet public-events deadlines.

Appendix 1

FACIAL GEOMETRY COMPARISONS

(When good photos with frontal views are available on both subject and previous life, one for each can be used for geometric-ratio comparisons. When only portraits, images of sculptures, or poor quality photographs are available, measure two or three and then average the ratios.)

Subject — Measurements in Millimeters and Ratios as Indicated

1. Eyes width (from outer eye corners).		Ratio #1a		Ratio #1b		Ratio #1c	Avg #1 Ratios
2. Eye socket height (from top edge to bottom).							
3. Nose length (from lowest point between eyes to tip).		Ratio #2a		Ratio #2b		Ratio #2c	Avg #2 Ratios
4. Nose bridge width (from inner eye corners).							
5. Span of cheekbones (from highest point on each side).		Ratio #3a		Ratio #3b		Ratio #3c	Avg #3 Ratios
6. Lower face length (lower edge of nose to point of chin).							

Previous Life

1. Eyes width (from outer eye corners).		Ratio #1a		Ratio #1b		Ratio #1c	Avg #1 Ratios
2. Eye socket height (from top edge to bottom).							
3. Nose length (from lowest point between eyes to tip).		Ratio #2a		Ratio #2b		Ratio #2c	Avg #2 Ratios
4. Nose bridge width (from inner eye corners).							
5. Span of cheekbones (from highest point on each side).		Ratio #3a		Ratio #3b		Ratio #3c	Avg #3 Ratios
6. Lower face length (lower edge of nose to point of chin).							

Ratio #1: Divide the distance in Line 2 by the distance in Line 1.
Ratio #2: Divide the distance in Line 4 by the distance in Line 3.
Ratio #3: Divide the distance in Line 6 by the distance in Line 5.

Averages of ratios: Add the ratios for #'s a, b, and c. Divide total by the # of ratios.

Variances between Subject and Previous Life (used with single measure or averages) are the differences between their respective ratios.

Ratio #1 Variance	Ratio #2 Variance	Ratio #3 Variance

Appendix 2

CEREBROTYPE FACTOR RATING SCALE:
Each continuum reflects a range of degrees between two extremes of a given characteristic. There is no right or wrong answer. All of us fall somewhere on the spectrum from 1 to 5. Insert initials at the most appropriate respective points for subject and previous personality.

This factor describes an individual's COGNITIVE modes. How does the person gather data, select what she uses, analize it, relate it to existing knowledge, and draw conclusions? How does he handle questions or issues, including a lack of information, and make decisions?

Emotional				Rational
1	2	3	4	5

Traditional				Experimental
1	2	3	4	5

Reactive				Disciplined
1	2	3	4	5

Global				Particular
1	2	3	4	5

Reflective				Impulsive
1	2	3	4	5

Scoring Guidelines: Remember this process is not an exact science. It is highly unlikely one can obtain equally reliable IQ scores on S and PP. While such measures would be useful if they were available, most cases must depend on the biographical data available through interviews, document reviews or indirect measures of mental styles and capacities. On the basis of the best available data, select the point on each continuum that shows where the S and PP would likely fit. Since there is no right or wrong answer, the rating process seeks to determine whether the two personalities have more similarities than one might find in two people picked at random.

Factor Rating Process: Based on a representative sample of relevant information on each lifetime, circle (adding notes for clarification when required) the number on each scale above that most accurately fits each personality. Use initials to identify which score applies to each person.

Factor Scoring: The same number or two adjacent numbers may be considered a correspondence. A gap between the selected numbers on each characteristic indicates the lack of correspondence. Four or five correspondences may indicate a strong likelihood of a past-life match. Three correspondences may indicate a possible match. Only 1 to 2 correspondences suggests an unlikely match.

Check one: High-level (4-5) = ____(3) Mid-level (3) = ____(2) Low-level (1-2)=____(1)

Appendix 3

EGOTYPE FACTOR RATING SCALE:
This factor deals with an individual's EMOTIONAL state. How does he react to daily life and his environment, particularly in stressful work, social or personal situations ? What is her level of energy when involved in a normal routine. What is one's default emotional state?

Each continuum reflects a range of degrees between two extremes of a given characteristic. There is no right or wrong answer. All of us fall somewhere on the spectrum from 1 to 5. Insert initials at the most appropriate respective points for both S and PP.

Cool				Warm
1	2	3	4	5
Confident				Worried
1	2	3	4	5
Depressed				Manic
1	2	3	4	5
Anxious				Calm
1	2	3	4	5
Optimist				Pessimist
1	2	3	4	5

Scoring guidelines: Remember this process is not an exact science. Even though the specific situations you have documented may vary, examine how each person has responded to them. Then pick out point on each scale that best places the person's emotional stance. With no right or wrong answer, the researcher does her best to infer from biographical data what the person's aproach has been in a variety of situations. To the extent one discerns long terms patterns, the more useful the data.

The goal is to determine whether the two personalities have more similarities than one might find between two people picked at random from the local shopping mall. If you find four or five similarities then the likelihood of a match is increased. One or two similarities can be easily attributed to chance.

To establish a complete file that can be reviewed by a third party, include copies of the biographical data used or quotes with the source references. The ability of a third party to verify the compiler's conclusions enhances the reliability of the scores and the level of confidence assigned to them.

Egotype Factor Scores: Circle the number on each continuum that most accurately fits each personality and include the S and PP initials within the box. Add other terms if needed to better describe both personalities. Four or five correspondences (including ajdacent scores) indicate confidence in a match. Three indicate only a possible match. One or two may be only a chance correspondence.
<u>Check one</u>: High-level (4-5) = ____(3) Mid-level (3) = ____(2) Low-level (1-2)=____(1)

Appendix 4

PERSONATYPE FACTOR RATING SCALE:

This factor focuses on the quality of a person's INTERPERSONAL style. How one engages others is a central aspect of the personality developed over time. It involves our prefudices and habitual ways of responding to people in a variety of circumstances

Each continuum reflects a range of degrees between two extremes of a given trait. The numbers in-between reflect more or less of a tendency. All of us fall somewhere on the spectrum from 1 to 5. Insert initials at the most appropriate point for S and PP.

Timid				Uninhibited
1	2	3	4	5
Aggressive				Submissive
1	2	3	4	5
Dependent				Independent
1	2	3	4	5
Introverted				Extraverted
1	2	3	4	5
Trusting				Skeptical
1	2	3	4	5

<u>Scoring guidelines</u>: After you have selected biographical information about both personalities, look for descriptions of behaviors or situations that generally reflect the two personalities. For each of the five scales choose the number most relevant to each personality. Keep in mind that the goal is not to find exactly the same situations in both lives, but to determine to what extent the two personalities have the same underlying interpersonal styles.

<u>Scoring Process</u>: After you have decided which point on each scale most accurately fits each personality, circle and make any appropriate notes in or near the blocks. Four or five correspondences (including adjacent ratings) may indicate the likelihood of a match. Three may indicate a possible correspondence. Only a 1 to 2 suggests an unlikely match.

<u>Note Overall Score</u>: High-level(4 or 5) = 3 ___ Mid-level(3) = 2___ Low-level(1 to 2) = 1___

<u>Reminder</u>: Remember when working on each page of this form that regardless of the score for a particular set of factors, it alone does not validate the alleged past-life match. Only when taken in the context of all other data in the case does it buttress or weaken the hypothesis that the subject is the same psychoplasm who also incarnated in the previous life under study.

Appendix 5

PERFORMATYPE FACTOR RATING SCALE

This scale identifies the VOCATIONAL traits that determine the areas in which people choose to devote their creative energies. They may or may not reflect the work a person has to do to earn a living. The broad categories are based on a set of underlying values, skills and interests. The jobs listed are only examples that illustrate a general cluster. Note the S and PP interests in the boxes of most relevance.

1. **ARTISTIC** - Non-conforming, original, independent, creative and, sometimes, chaotic individuals. They value beauty and imagination through self-expression, art and communication.

Musician	Dancer	Author/Poet	Painter	Art Therapist	Actor

2. **CONVENTIONAL** - Values precision, attention to detail, and orderly processes. Likes organizational stability, procedures, efficiency and status.

Retail	Engineer	Banker	Administrator	Statistician	Editor

3. **ENTERPRISING** - Competition, leading, persuading, selling, dominating, and promoting. It involves risk taking and self-promotion.

Politician	Salesman	Publisher	Consulting	Journalism	Businessman

4. **INVESTIGATIVE** - Curiosity and learning are the core of this factor, with its focus on systematic or scientific information gathering, analysis, and theory building.

Economist	Psychiatrist	Mathematician	Professor	Surgeon	Scientist

5. **REALISTIC** - Working with hands, things, tools, or machines. Practical, physical and mechanically inclined. Values tradition and common sense. Good physical coordination with ingenuity and dexterity.

Architect	Veterinarian	Mechanic	Chef	Pilot	Athlete

6. **SOCIAL** - Values cooperation, generosity, and service to others. Focus on teamwork and community. Requires people skills, verbal ability, listening and understanding.

Nurse	Teacher	Minister	Psychologist	Physician	Trainer

Scoring Guidelines: Document specific activities or events in both lives that indicate at least each subject's primary, secondary, and tertiary occupational interests and strengths. (If evidence suggests more correspondences, note them as well.) Describe each with one or two words in appropriate box.

Scores. If you find three or more significant matches, they suggest a high level of confidence (3). Two matches indicate a mid-level of confidence (2). One or no matches should be considered a score of (0).

Endnotes

Introduction

1. Ian Stevenson, *Where Reincarnation and Biology Intersect* (Westport: Praeger, 1997), 112–13.

2. The personal relationships involved in the Charlotte, Kelly, and James-II cases required diligent efforts by the parties to remain objective during the research process. Their personal connections may also have influenced the tentative nature of the past-life identifications made for Charlotte and Kelly by Ahtun Re. Before Charlotte became involved in past-life research, she and James-II had had a five-month, long-distance, dating relationship. The identification of a past-life connection for Charlotte came when James-II invited her to a conference on reincarnation. Walter Semkiw, as convenor of the meeting, asked Ahtun Re about Charlotte's past lives. At that point Ahtun Re identified Charlotte as the reincarnation of Dolley Madison.

A year later, James-II invited his then-wife Kelly to a follow-up meeting of the same group. Semkiw asked Ahtun Re for a past-life identification for Kelly. Ahtun Re responded that Kelly and Charlotte were splits of the Dolley soul. Prior to both channeled sessions, Semkiw, Ryerson, and Ahtun Re knew of the previously-identified, independent case for James-II as a possible reincarnation of James Madison. It is not clear what impact this knowledge had on extradimensional past-life identifications for the two women associated in this life with James-II.

From the first meeting, Charlotte was interested "in figuring out the truth" about a possible lifetime during the Revolutionary Era. On her behalf, Semkiw sought clarification from Ahtun Re who reportedly suggested the soul-split phenomenon meant she was a "personality aspect" of the Dolley "core identity." At one point she was informed her present life experience was as a "part of a parallel life with Kelly that was once Dolley."

During a past-life reading by John Jones, known as a "karmic healer," Charlotte asked him if he saw a lifetime for her during the Founding Era. He supported her understanding that she had been Dolley Madison. (For information on John Jones see www.karmichealer.com.)

In another reading, Charlotte asked Candace Smith (www.campchesterfield.net) if she saw a past-life for her from the mid-1700s to the mid-1800s. Candace stated she saw Charlotte as a woman extremely close to Thomas Jefferson and to someone else who was close to Jefferson, and said that she was an inspiration to them. Charlotte took that to mean a lifetime as Dolley.

Seeking further understanding of the soul-split concept, Charlotte asked psychologist and hypnotherapist Linda Backman for her view of soul-splits. See discussion on page 200. (Backman can be contacted at www.ravenheartcenter.com.)

That these interpersonal and extradimensional experiences influenced the attitudes about and the reporting of information in these cases is highly likely. Given that situation, Charlotte, Kelly, and James-II attempted to test and re-test their own assumptions, through personal interviews and discussions, to be as objective as possible during participation in the project.

Chapter One

1. Jim B. Tucker, *Life Before Life* (New York: St. Martin's Press, 2005), 82.

2. Tucker, 123.

3. Jesse Bering and David Bjorklund, discussed by Paul Bloom, *The Atlantic Monthly*, December 2005.

4. Paul Bloom and Deena Skolnich Weisberg, "Why Do Some People Resist Science?" 9 May 2007, <http://www.edge.org/documents/archive/edge211.html#bloom2> (7 December 2007).

5. Tucker, 134.

Chapter Two

1. Wayne Peterson, interview by Paul Von Ward, Founding Mystics Meeting, 3 July 2005.

2. Ming Liu. "Chomsky and Knowledge of Language," Twentieth World Congress of Philosophy, 15 August 1998 < www.bu.edu/wcp/Papers/Lang/LangLiu2.htm> (7 December 2007).

3. Kim Edward Adams and Dee Loecher, interview by Paul Von Ward, Las Vegas, Nevada, 1 July 2007, <www.spirituntld.com>.

4. Jeffrey J. Keene, *Someone Else's Yesterday* (Nevada City, CA: Blue Dolphin Publishing, 2003), 22.

5. Keene, 59.

Chapter Three

1. Brian O'Leary, *Exploring Inner and Outer Space* (Berkeley, CA: North Atlantic Books, 1989), 132–33.
2. Brian O'Leary, *Miracle in the Void* (Kihei, HI: Kamapua'a Press, 1996), 146.

Chapter Four

1. Jonathan Kolber, email to author, 23 August 2006.
2. Roger J. Woolger, *Other Lives, Other Selves*. New York: Doubleday, 1987.
3. Bruce Goldberg, *Past Lives, Future Lives*. New York: Ballantine Books, 1982. Edith Fiore, *You Have Been Here Before*. New York: Ballantine Books, 1985. Brian Weiss, *Many Lives, Many Masters*. New York: Fireside Books, 1988. Adrian Finkelstein, *Marilyn Monroe Returns*. Charlottesville, VA: Hampton Roads Publishing, 2006.
4. Carol Bowman, *Children's Past Lives* (New York: Bantam Books, 1997), 192.

Chapter Five

1. Tucker, 9–10.
2. Tucker, 7–8.
3. Stevenson, 165.
4. Keene, 8.
5. Stevenson, 162.

Chapter Seven

1. David C. McClelland, *Power: The Inner Experience*. New York: Halsted Press, 1975.
2. As a result of the author's work with McClelland, his official colleagues in Washington agreed to seek McClelland's assistance with several government initiatives. In the mid-70s, the Harvard-USG team developed several programs with the aim of increasing professional competence in public-service professionals.
3. Paul Von Ward, *Our Solarian Legacy: Multidimensional Humans in a Self-Learning Universe*. Charlottesville, VA: Hampton Roads Publishing, 2001.
4. Ian Stevenson, *Twenty Cases Suggestive of Reincarnation*. Charlottesville, VA: University of Virginia Press, 1974.
5. Walter Semkiw, *Return of the Revolutionaries*. Charlottesville, VA: Hampton Roads Publishing, 2003.

6. The author is grateful to Semkiw for having exposed such biographical data in his presentations at the July 2005 gathering of the Founding Mystics. While the subjects of some cases described in this book were present at that meeting, the "reincarnation experiment" and the work presented in *The Soul Genome* are independent of the Founding Mystics and the Institute for the Integration of Science, Intuition, and Spirit.

Chapter Eight

1. Laurence Krauss, "The Physics of Nonsense," 1 August 2007, Newsletter <http://www.eSkeptic.com.>

2. Michael Talbot, *Holographic Universe*. New York: HarperCollins, 1991.

3. G.E. Morfill, et al., "From Plasma Crystals and Helical Structures toward Inorganic Living Matter." *New Journal of Physics* 9, no 263 (August 2007).

4. Cleve Backster, *Primary Perception: Biocommunication with Plants, Living Foods, and Human Cells*. Anza, CA: White Rose Millennium Press, 2003.

5. William Tiller, "Some Science Adventures with Real Magic: Theoretical Explanation of Experimental Data," 22 April 2006, oral presentation, International Conference on Science and Consciousness in Santa Fe, NM.

6. Dean Radin, *The Conscious Universe*. San Francisco: HarperSanFrancisco, 1997; and *Entangled Minds*. New York: Paraview Pocket Books, 2006.

7. Von Ward (2001).

8. Gary E. Schwartz, *The Afterlife Experiments*. New York: Pocket Books, 2002.

9. See Bibliography for suggested texts on this research.

10. Monroe Institute web site <http://www.monroeinstitute.com/program.php?program_id=8>. Overview of the Development of Lucid Dream Research in Germany <http://gestalttheory.net/archive/thol_lucid2.html>. Lyn Buchanan, Department of Defense-trained remote viewer, web site <http://www.crviewer.com/>.

11. Kevin Ryerson and Stephanie Harolde, *Spirit Communication*. New York: Bantam Books, 1989.

12. Jacob D. Bekenstein. "Information in the Holographic Universe," *Scientific American* 289 no 2 (August 2003), 58–65.

13. Inomata previewed this concept for Paul Von Ward in a TV interview done at the Institute for New Energy Conference in Denver,

Colorado in May 1994. Also see Jakob Hasselberger December 1996 article at <http://homepages.ihug.co.nz/~sai/newparsci.htm> (28 December 2007).

14. Thomas Jefferson (letter to William Johnson, 12 June 1823), *The Writings of Thomas Jefferson*, Memorial Edition. Lipscomb and Bergh, eds., vol. 15, 450.

15. A review of metaphysical and paranormal explanations for evidence associated with reincarnation is available to the reader on the Reincarnation Experiment web site <htttp://www.reincarnationexperiment.org>.

Chapter Nine

1. Stevenson (1997), 183.

2. Derived from the Greek roots *psych* and *plasm*, the word psychoplasm indicates a primordial combination of the psychical and the physical.

3. Alex Mauron "Is the Genome the Secular Equivalent of the Soul?" *Science Magazine* (2 February 2001): 831-2.

4. Savely Savva, Editor, *Life and Mind: In Search of the Physical Basis*. Oxford: Trafford Publishing, 2007.

5. Bruce Lipton, *Biology of Belief* (Santa Rosa, CA: Elite Books, 2005), 190.

6. Stevenson (1997), 119–20.

7. Neil Whitehead, "The Importance of Twin Studies" at <http://www.narth.com/docs/whitehead2.html.> He writes identical twin studies show that neither genetic nor family factors overwhelm tendencies to individuation.

8. *Research in Review*, Florida State University, Winter 2006, 41.

9. See basic definition of epigenetics at <http://en.wikipedia.org/wiki/Epigenetics> (28 December 2007). Epigenetic features are inherited when cells divide despite a lack of change in the DNA sequence itself.

10. Stevenson (1997), 182–83.

11. <http://homepages.ihug.co.nz/~sai/DNAPhantom.htm> 19 March 2002. In theoretical physics the worm hole acts as a tunnel-like connection that transmits information outside of the limitations of time and space. (28 December 2007).

12. Adrian Finkelstein, *Marilyn Monroe Returns* (Chalottesville, VA: Hampton Roads Publishing, 2006), 95.

13. Jeffrey Mishlove and Brendan Engen. Draft paper "Archetypal Synchronistic Resonance: A New Theory on Paranormal Experience" shared with author August 2007.

14. The SAR concept is explained in more detail on the Reincarnation Experiment web site.

15. Ervin Laszlo, *Science and the Akashic Field: An Integral Theory of Everything* (Rochester, VT: Inner Traditions, 2004), 70.

Chapter Ten

1. Stevenson (1997), 165.
2. Semkiw, 33.
3. Finkelstein, 162–63.
4. For an introduction to facial recognition systems, go to <http://computer.howstuffworks.com/facial-recognition.htm.> (28 December 2007).
5. April 2005 <http://web.telia.com/~u57013916/Edlinger%20Mozart.htm> (28 December 2007).
6. Semkiw, 4, 33, 44, & 48.
7. Bio-Identification: Frequently Asked Questions, 10 June 2007 <http://www.bromba.com/faq/biofaqe.htm#Merkmale> (28 December 2007).
8. Sheryl Ubelacker, "Each Person's Genetic Make-up Differs Far More Than Previously Known," *Canadian Press*, 22 November 2006. Alexander Bimelbrant, et al., "Widespread Monoallelic Expression on Human Autosomes," *Science*, 16 November 2007, Vol. 318, no. 5853, 1136–1140.
9. Samuel Levy, et al., "The Diploid Genome Sequence of an Individual Human." Public Library of Science: Biology, 8 May 2007 <PLoS Biol 5(10): e254 doi:10.1371/journal.pbio.0050254> (28 December 2007).
10. " Genetic On-Off Switches Pinpointed in Human Genome," National Science Foundation Press Release 05-108, 29 June 2005 <www.nsf.gov/news/news_summ.jsp?cntn_id=104282> (28 December 2007).

Chapter Eleven

1. Raymond B. Cattell, *The Scientific Analysis of Personality*. Chicago: Aldine Publishing, 1966, and Stella Chess and Alexander Thomas, *Temperament and Development*. New York: Bruner/Mazel, 1977.
2. Stephan A. Schwartz is the author of *Opening to the Infinite* (Nemoseen Media, 2007).
3. Private subscription of Schwartz Report, May 15, 2007.
4. Roger Lewin, "Is Your Brain Really Necessary?" *Science*, 12 December 1980, Vol. 210, 1232.

5. Stevenson (1974), 66-7. Note: James-II recalled inexplicable early childhood criticisms of his biological family's table manners and their refusal to eat fresh vegetables and "fancy" dishes. Later insight into James and Dolley's efforts to create a cosmopolitan table offered James-II a possible explanation for his early behavior.

6. Stevenson (1974), 53–56.

7. Stevenson, (1974), 92.

8. Stevenson (1974), 41 & 45.

9. For more background information, see the University of Kansas Career Counselling and Planning Services web site at <http://www.caps.ku.edu/career/tests.shtml> (28 December 2007).

10. John L Holland, *Making Vocational Choices: A Theory of Vocational Personalities and Work Environments*. Davis, CA: Psychological Assessment Resources Inc., 1997.

11. Thomas Jefferson, "Answers to de Meusnier Questions, 1786," *The Writings of Thomas Jefferson*, Lipscomb and Bergh, eds. 17:8.

Chapter Twelve

1. Finkelstein, 92.

2. Finkelstein, 95–6.

3. Bowman, 8.

4. Debra Rosenberg, "Rethinking Gender," *Newsweek*, 21 May 2007, 57.

5. James H. Kent, *Past Life Memories as a Confederate Soldier*. Huntsville, AR: Ozark Mountain Publishers, 2003.

6. Similar affinity cases of present-day neurosurgeon Norman Shealy and 19th-century English physician John Elliotson and present-day psychologist Jeffrey Mishlove and nineteenth-century psychologist William James are presented in *Return of the Revolutionaries* and *Born Again* by Walter Semkiw.

7. Marcia Schafer, *Confessions of an Intergalactic Anthropologist* (Phoenix, AZ: Cosmic Destiny Press, 1999), 60 & 111.

8. During the experiment it became clear that some people believe that an extradimensional source should be used as the confirmation of a past-life identification. In other words, they trust the "spiritual" realm more than their own evidence. This illustrates the power of metaphysical worldviews.

Chapter Thirteen

1. Ralph Ketcham, *James Madison: A Biography* (Charlottesville, VA: University of Virginia Press, 1990), 35.
2. Finkelstein, 44.

Chapter Fourteen

1. Ketcham, 470–73.
2. Robert Allen Rutland, *James Madison: The Founding Father* (Columbia, MO: University of Missouri Press, 1987), 19 & 30.
3. Citation available in author's files.
4. Citation available in author's files.
5. Citation available in author's files.
6. Dead Sociologists Society Index <http://media.pfeiffer.edu/lridener/DSS/#sorokin> (28 December 2007).
7. Finkelstein, 168–69.

Chapter Fifteen

1. Finkelstein, 54.
2. Finkelstein, 149 & 172.
3. Walter Semkiw, *Born Again* (New Delhi: Ritana Books, 2006), 127.
4. Keene, 35–6.
5. Keene, 37.
6. Keene, 2.

Chapter Sixteen

1. Conover Hunt-Jones, *Dolley and the "great little Madison"* (Washington, D.C.: American Institute of Architects Foundation, 1977), 23.
2. Hunt-Jones, 23.
3. Hunt-Jones, 35.
4. Lorin Kee Web Site: <http://www.thegnosticoracle.com.> (28 December 2007).
5. Finkelstein, 179.

Chapter Seventeen

1. Anne Frank, *The Diary of a Young Girl*. New York: Bantam, 1993.
2. Barbro Karlen, *And the Wolves Howled*. London: Clairview, 2000.
3. Semkiw (2006), 134.

Chapter Eighteen

1. Semkiw (2003), 37.
2. Semkiw (2006), 132.
3. Keene, 27.
4. Rupert Sheldrake, *The Sense of being Stared At: And Other Aspects of the Extended Mind.* London: Arrow Books, 2004; and Dean Radin (2006).
5. The notion that a true physics would include humans and their consciousness in a coherent picture of the world was promoted by Pierre Teilhard de Chardin in his book *The Phenomenon of Man*, New York: Harper, 1976.
6. See discussion of methodology by Lea Winerman, "A Second Look at Twin Studies," *Behavioral Genetics* 35, no. 4 (April 2004): 46.
7. Semkiw PowerPoint presentation at Founding Mystics meeting, 4 July 2005.
8. Keene, 79.

Chapter Nineteen

1. Semkiw (2003) 178-82.
2. Permutations of the soul-twin concept can be found in Stevenson (1997), 161–2, 164, 168, & 169–70.
3. Tucker, 60, 128 & 131.
4. Monozygotic or identical twins result from the division of a single fertilized egg between 1 and 14 days after conception. They share the same genes, with minimal differences resulting from mutations—changes in the cell that are not corrected prior to or during its replication.
5. Proponents of the soul-split idea avoid this conundrum by saying the two individuals who were split from the same soul have different fragments of the original (in order to have different experiences). That begs the questions of where the "non-split" elements that make up the rest of the whole present incarnation came from. Did they come from even another soul splitting itself? If so, where did the other part of that soul go? Where did that incarnation get its other fragments? The resulting unending dilemma denies the reincarnation of any integral souls and depends on the notion of infinite, parallel realities acting outside natural processes susceptible to evaluation.
6. Kevin Ryerson, e-mail to author, 5 February 2007.

Bibliography

Backster, Cleve. *Primary Perception: Biocommunication with Plants, Living Foods, and Human Cells.* Anza, CA: White Rose Millennium Press, 2003.
Bowman, Carol. *Children's Past Lives.* New York: Bantam Books, 1997.
Cattell, Raymond B. *The Scientific Analysis of Personality.* Chicago: Aldine Publishing, 1966.
Chess, Stella and Alexander Thomas. *Temperament and Development.* New York: Bruner/Mazel, 1977.
Finkelstein, Adrian. *Marilyn Monroe Returns.* Charlottesville, VA: Hampton Roads Publishing, 2006.
Fiore, Edith. *You Have Been Here Before.* New York: Ballantine Books, 1985.
Frank, Anne. *The Diary of a Young Girl.* New York: Bantam, 1993.
Goldberg, Bruce. *Past Lives, Future Lives.* New York: Ballantine Books, 1982.
Holland, John L. *Making Vocational Choices: A Theory of Vocational Personalities and Work Environments.* Davis, CA: Psychological Assessment Resources Inc., 1997.
Hunt-Jones, Conover. *Dolley and the "great little Madison."* Washington, D.C.: American Institute of Architects Foundation, 1977.
Barbro Karlen, Barbro. *And the Wolves Howled.* London: Clairview, 2000.
Keene, Jeffrey J. *Someone Else's Yesterday.* Nevada City, CA: Blue Dolphin Publishing, 2003.
Kent, James H. *Past Life Memories as a Confederate Soldier.* Huntsville, AR: Ozark Mountain Publishers, 2003.
Ketcham, Ralph. *James Madison: A Biography* (Charlottesville, VA: University of Virginia Press, 1990),
Laszlo, Ervin. *Science and the Akashic Field: An Integral Theory of Everything.* Rochester, VT: Inner Traditions, 2004.
Lipton, Bruce. *Biology of Belief.* Santa Rosa, CA: Elite Books, 2005.
McClelland, David C. *Power: The Inner Experience.* New York: Halsted Press, 1975.
O'Leary, Brian. *Exploring Inner and Outer Space.* Berkeley, CA: North Atlantic Books, 1989.

O'Leary, Brian. *Miracle in the Void*. Kihei, HI: Kamapua'a Press, 1996.

Radin, Dean. *The Conscious Universe*. San Francisco: HarperSanFrancisco, 1997.

Radin, Dean. *Entangled Minds: Extrasensory Experiences in a Quantum Reality*. New York: Paraview Pocket Books, 2006.

Rutland, Robert Allen. *James Madison: The Founding Father*. Columbia, MO: University of Missouri Press, 1987,

Ryerson, Kevin and Stephanie Harolde. *Spirit Communication*. New York: Bantam Books, 1989.

Savva, Savely (Ed). *Life and Mind: In Search of the Physical Basis*. Oxford: Trafford Publishing, 2007.

Schafer, Marcia. *Confessions of an Intergalactic Anthropologist*. Phoenix, AZ: Cosmic Destiny Press, 1999.

Semkiw, Walter. *Return of the Revolutionaries*. Charlottesville, VA: Hampton Roads Publishing, 2003.

Semkiw, Walter. *Born Again*. New Delhi: Ritana Books, 2006.

Schwartz, Gary E. *The Afterlife Experiments*. New York: Pocket Books, 2002.

Schwartz, Stephan A. *Opening to the Infinite*. Nemoseen Media, 2007.

Sheldrake, Rupert. *The Sense of Being Stared At: And Other Aspects of the Extended Mind*. London: Arrow Books, 2004.

Stevenson, Ian. *Twenty Cases Suggestive of Reincarnation*. Charlottesville, VA: University of Virginia Press, 1974.

Stevenson, Ian. *Where Reincarnation and Biology Intersect*. Westport, CT: Praeger, 1997.

Talbot, Michael. *Holographic Universe*. New York: HarperCollins, 1991.

Teilhard de Chardin, Pierre. *The Phenomenon of Man*, New York: Harper, 1976.

Tucker, Jim B. *Life Before Life*. New York: St. Martin's Press, 2005.

Von Ward, Paul. *Our Solarian Legacy: Multidimensional Humans in a Self-Learning Universe*. Charlottesville, VA: Hampton Roads Publishing, 2001.

Weiss, Brian. *Many Lives, Many Masters*. New York: Fireside Books, 1988.

Woolger, Roger J. *Other Lives, Other Selves*. New York: Doubleday, 1987.

Index

Adams, John, 130, 191
Adams, Kim, 23
Affinity cases, 113
The Afterlife Experiments, 64, 118
Akashic records, 57, 116
Alexander, Ken, 8, 26–8, 116, 120, 130, 132, 145, 168, 174–5
Almeida, Monica, 13
Alternative hyotheses, xii, 80–1
And the Wolves Howled, 9
Aristotle, 2
Aspirations, 29
Atlantic Monthly, 17
Atun Re, 9, 77, 117
Atwater, P.M.H., 64
Avocation, 178

Backster, Cleve, 63
Baldwin, James M., 69
A Beautiful Mind, 14
Bekenstein, Jacob, 65
Biocommunication, 63
Biofield, 71, 74–5
Biographical material, 154
Biometrics, 86–9, 125
Biometric scoring, 119
Birthmarks, 38
Black hole, x, 3
Bloom, Paul, 17
Body types, 40, 74, 89, 90, 129–32
Bowman, Carol, 36, 111
Brain/mind, 97

Cathartic release, 34
Cayce, Edgar, 57, 116

Cerebrotype, 7–9, 96–9, 137–48
Channel, 116
de Chardin, Teilhard, 57
Charlotte, 4, 24, 127–8, 130, 134, 138–9, 156–8, 162, 165, 178–9, 184, 199–200
Chase, 111–2
Children's Past Lives, 36
Chomsky, Noam, 22, 171
Cognitive traits, comparisons, 139, 146
Cohorts (soul), 194–205
Coincidences, 181, 183, 186–93
Collective unconscious, 57
Confessions of an Intergalactic Anthropologist, 116
The Conscious Universe, 64
Conservation of consciousness, 66
Conservation of energy, 66
Correspondences, 115
Cryptomnesia, 25

Dalai Lama, ix, xi
Deformities, 8
Delbruck, Max, 69
Denver, John, 9, 85, 127, 175–6, 190, 200–2
Different scores, 158
Disincarnation, 63
Doppelganger, 37
DNA, 63, 70, 73–6, 83, 91–3, 183, 198
Dreams, 26, 28

INDEX

Ear form, 90, 132
Ectomorph, 39, 130, 201
Ego Factor Scale, 150, 157
Egotype, 79, 96, 100–1, 140–59
Einstein, Albert, 3, 184
Elliotson, John, 176
Endomorphs, 130
Entangled Minds, 64, 185
Entanglement, 184–5
Evaluation forms, 115
Exploring Inner and Outer Space, 9
Extradimensional sources, beings, xi, 115–6, 186

Facial features, geometry, and architecture, 84–9, 125–9, 201
Falsifiability, x
Fingerprints, 90
Finkelstein, Adrian, 35, 39, 77, 89, 110–11, 145
Five factors, 78, 180
Frank, Anne, 9, 20, 170
Free association, 34
Freud, Sigmund, 34, 57

Gad/Gauguin, Mette, 5, 23–4, 40–1, 99, 134, 151, 166–7, 172, 194
Galileo, Galilei, 2
Gariaev, Peter, 76
Gauguin, Paul, 5, 24, 40–1, 99, 105–6, 110, 136, 151, 166–7, 172, 194
Genetic mutations, 31
Genotype, 78, 84, 90–3, 129, 135, 203
Gestation, 75
Glossolalia, 21
Gordon, John B., 5, 25, 31, 38–9, 107, 130–2, 152–3, 177, 187

Grant, Joan, 19
Grey Eagle, John, 23
Greenberg, Jay, 13

Habits, 186
Hallucination, 33
Hands and fingers, 90, 133
Harvard University, 56, 139
Heraclitus, 181
Holland and Strong, 106
Holographic transfer, 60
The Holographic Universe, 62
Houle, David, 74
Hypnosis, 111
Hypothesis, x

Ibrahim, 31
Identical twins, 74
Imad, 22
Indigo children, 17
Inomata, Shiuji, 65
Integral Model, x, 61, 82, 115, 119, 121, 150, 158, 169
Integral universe, 66
International Association for New Science, 1
Intuitive, 17, 119

James-II, 7, 39, 40, 43–52, 120, 126–7, 132–5, 140–3, 162–4, 173–4, 176–7, 179, 182–3, 187–93
Jasbir, 31, 101, 171
Jatav, Lekh Pal, 73
Jefferson, Thomas, 4, 6, 108, 130, 147, 180, 189, 191–2
Johnson, Samuel, 6, 20, 133–4
Journey of Souls, 26
Jung, Carl, 34, 57, 181

Karma, 203
Karlen, Barbro, 9, 15, 20–1, 120, 170, 189
Kee, Lorin, 8, 113–4, 120, 127, 129, 130, 132, 143–4, 153, 157, 167, 172–3, 178, 188
Keene, Jeffrey J., 6, 25, 31, 38–9, 107, 120, 130–2, 135, 152–3, 177, 182, 187, 195
Kelly, 4, 103–4, 127-8,130, 134, 137-9, 156–7, 159, 165, 169, 178–9, 182, 184, 187–9, 192–3
Kent, James H., 9, 25, 28–9, 77, 113
Kern, Steve, 9, 85, 127, 175–6, 190, 200–2
Ketcham, Ralph, 139
Kolb, Susan, 30
Kolber, Jonathan, 33
Krauss, Lawrence, 62

Laird, Sherrie, 7, 88, 110–11, 120, 133–5, 145, 149, 157, 168, 177, 184
Lamarck, Jean-Baptiste, 75
Laszlo, Ervin, 81
Lemke, Leslie, 15
Lewin, Roger, 97
Life Before Life, 31, 135
Life-path applications, 180
Linguists, 22
Lipton, Bruce, 71
Loecher, Dee, 23
Long, Jeffery, 64
Look-alikes, 85, 127
Lorber, John, 98
Lucy (Payne Todd Washington), 154–5, 157–8, 160–2, 164–5, 184

Madison, Dolley, 4, 24, 103–4, 127–8, 130, 134, 137–9, 156–8, 159, 164–5, 169, 173–4, 178–9, 182, 187–9
Madison, James, 4, 7, 39, 40, 42–52, 126–7, 130, 132–5, 139–43, 154, 162–4, 176–7, 179, 182–3, 187–93
Marilyn Monroe Returns, 39
Mauron, Alex, 71
McClelland, David, 56, 96, 139
McConnell, John and William, 135
Mesomorph, 40, 130, 201
Meta-analysis, 77
Methodology, 155
Mettler, Darlene, 20, 109
Mettler, George B., 2, 20, 133, 171
Mishlove, Jeffrey, 80
Monroe, James, 191
Monroe Institute, 65
Monroe, Marilyn, 7, 88, 133–4, 149, 157, 177, 184
Moshay, Michelle, 5, 23–4, 40–1, 99, 113, 120, 134–5, 151, 166–7, 172, 194
Motivation theory, 56
Mozart, W. A., 14

NASA, 63
Newton, Isaac, 3
Newton, Michael, 26
Nash, John, 15
Natural philosophy, 2
Natural science, 2
Nelson, Willie, 16
Neuroscience and neuroscientists, 18
New Journal of Physics, 63
Nonhuman sources, 117
Non-locality, 184–5
Noosphere, 57

OBE's, 65
Occam's razor, 67
Odors, 90, 93
Old souls, 17
O'Leary, Brian, 9, 29, 30, 136, 147
Opening to the Infinite, 97
Other Lives, Other Selves, 35
Our Solarian Legacy, 64, 68

Parmod, 31, 171
Parnia, Sam, 64
Past-life healing, 36, 33–6
Past-life identification, 168
Past-life regression, 36
Past-life therapy, 11, 34
Past Lives of a Confederate Soldier, 9
Pelley, Scott, 13
Performatype, 80, 96, 104–8, 170–80
Personality factors, 94
Personality theory, 95
Persona, 102, 161
Personatype, 79, 96, 101–4, 160–9
Peterson, Wayne, 20, 187
Phenotype, 78, 91, 125–36
Plato, 65, 196
Poetry, 137
Polygraph, 120
Poponin, Vladimir, 76
Precocious and precocity, 16, 17
Predictions, 147, 180
Probability theory, 3
Prodigy and prodigies, 11, 13, 14, 15
Psychophore, 70–1
Psychoplasm, xii, 70–82, 102, 136, 176, 183–4, 193, 202

Quick-starts, 15

Race, 39
Radin, Dean, 64, 183, 185
Ravi, 101
Reincarnation experiment, 55, 61, 203
Reincarnation hypothesis, x, xi, 126, 203, 205
Reliability, 118
Retroactive hints, 112
Return of the Revolutionaries, 39, 59
Roberts, Dalton, 16
Rutland, Robert, A. 139
Ryerson, Kevin, 9, 26, 65, 204

Savant, 15
Savva, Savely, 71
Schwartz, Gary, 64
Schwartz, Stephan A., 97
Schafer, Marcia, 116
Schrodinger, 184
Science and the Akashic Field, 81
Science Magazine, 71
Scolastico, Ron, 58
The Second Coming of Science, 9
Secondhand information, 115–6
Self-awareness, 94
Self-directed research, 121
Self-learning universe, 64, 68
Self-reporting, 110–1
Semkiw, Walter, 40, 59, 77, 89, 181, 187, 194
Shealy, Norman, 30, 176
Sheldrake, Rupert, 63, 183
Shelley, Percy, 19
Sorokin, Pitirim A., 7, 113–4, 127, 129, 130, 132, 143–4, 153, 157, 167, 173, 178, 188
Soul, x, xiii
Soul family, 196
Soul genome, 66, 73, 136, 203, 205

Soul mates, 196
Soul memory, 183
Soulprints, xii, 171
Soul siblings, 196
Soul-splits, 198–9
Sources and methods, 109–21
Special markings, 135
Steiner, June, 58
Stevenson, Ian, 1, 8, 37–8, 40, 59, 75, 101, 194, 197
Stigmata, 9
Strange attractors, 185
Special relativity, xiii
Sukla, 101, 107, 171
Superstring theory, 66
Swarnlata, 16, 22, 171
Swidersky, Tony, 8, 26–8, 117, 130, 132, 145, 168, 174–5
Synchronicities, 181, 193
Synchronistic Archetypal Resonance (SAR), 80

Talbot, Michael, 62
Tammet, Daniel, 15
Tao, Terrence, 13
Tastes, 186
Team approach, 120
Teekamp, Peter, 5, 16, 21, 23–5, 40–1, 99, 105–6, 110, 120, 135, 151, 166–7, 171–2, 194
Therapist, 169
Thirdhand data, 116
Third-party reports, 39
Third-party researcher, 169
Tiller, William, 64
TJ-?, 6, 108, 147, 180, 189
Translucent consciousness, 63
Tucker, Jim B., 1, 8, 16, 38, 59, 194, 197
Twenty Cases Suggestive of Reincarnation, 1

Twins analogy, cases, 185, 197
Two-life memory, 182

Unapparent connections, 181, 184
Unaquired knowledge, 9, 22
Unconscious, 57
Unexpected attitudes, 17
Unexpected choices, 23
University of Virginia, 1, 58, 76, 135, 189, 197
Unrealized dreams, 192–3
USS Airship Akron, 8, 27
U.S. National Research Council, 63

Variances, 125–6
Venter, Craig, 92
Virtuosos, 18
Vocations, 178
Voice, 90, 133–4

Weird life, 63
Weisberg, Deena S., 17
Where Reincarnation and Biology Intersect, 38
Wheeler, John A., 65
Whitehead, N.E., 74
Woolger, Roger, 34–5
Wormhole, 76
Wounds, 8, 39, 90
Writing, 137–8, 140–1

Xenoglossy, 22

Zeitlin, Malou, 19

About the Author

Paul Von Ward is an independent scholar who synthesizes the physical and life sciences with consciousness studies to create an interdisciplinary cosmology based on multiple epistemologies. His graduate-level research at Harvard (MPA) and Florida State University (MS) viewed the human personality and society from a psychological and developmental perspective. His work on reincarnation applies that background to the interface between the Standard Model of physics and the emerging theories of a multidimensional universe.

Prior to his research and paradigm-shifting books on the inner senses, subtle energies, and the impact on consciousness and culture of *Homo sapiens'* perceptions of advanced beings, Paul pursued a professional path including the Christian ministry, service as a U. S. naval officer, U. S. diplomat, and career foreign service officer. His book *Dismantling the Pyramid: Government by the People* influenced thinking about government reform in the 1980s. While serving as founder and CEO of the nonprofit Delphi International, he gained an international reputation in Soviet-American and Sino-American affairs as a citizen-diplomat. His column *Global Perspective* was published in Washington, D. C. from 1987 to 1995.

Paul's books *Gods, Genes, & Consciousness* (2004) and *Our Solarian Legacy: Multidimensional Humans in a Self-learning Universe* (2001) have established his reputation as a solidly-grounded researcher willing to experiment. He dares to point new directions for science's effort to expand human knowledge of the cosmos—in which we are "obvious parts of a self-evolving organism."

He lives in the woods of North Georgia and stays connected to the world through the Internet, his media events, and travels for his research and lectures. He can be reached through his web site www.vonward.com or the Reincarnation Experiment web site ww.reincarnationexperiment.org.

Printed in the United Kingdom by
Lightning Source UK Ltd., Milton Keynes
136733UK00002B/57/P